ANTHONY J. STUART
Castle Museum, Norwich

LIFE IN THE ICE AGE

SHIRE ARCHAEOLOGY

Cover photograph
Mammoths in a Norfolk landscape about 20,000 years ago
(Norfolk Museums Service, painting by N. Arber).

British Library Cataloguing in Publication Data available

Published by
SHIRE PUBLICATIONS LTD
Cromwell House, Church Street, Princes Risborough,
Aylesbury, Bucks HP17 9AJ, UK
Series Editor: James Dyer

ISBN 0 85263 929 5

First published 1988

Set in 11 point Times and printed in Great Britain by
C. I. Thomas & Sons (Haverfordwest) Ltd,
Press Buildings, Merlins Bridge, Haverfordwest, Dyfed.

Contents

Acknowledgements

I wish to express my gratitude to Kate Scott for reading an earlier draft of the manuscript, to Michael Bishop, Stephen Green, John Hallam, Mark Roberts, the Sedgwick Museum, Cambridge, Kate Scott and John Wymer for kindly supplying photographs, to John Crang for technical photographic help, to Anna Bartlett for the drawing of the giant deer skeleton and to Colin Shuttleworth and the Shropshire Museums Service for kindly supplying the photograph of the mammoth at Condover. The reconstructions of ice age scenes are by Nick Arber.

List of illustrations

1
Introduction

During the ice age, from perhaps 500,000 to 10,000 years ago, Britain was inhabited by peoples with old stone age (palaeolithic) cultures. They manufactured tools and weapons from natural materials, especially flint, wood and bone. Throughout the long span of prehistory man's existence was intimately bound up with animals and plants. Only at the end of the ice age did people, firstly in south-west Asia, begin to rely on sown crops and domestic animals for their main supplies of food. This new way of life did not reach Britain for another four or five thousand years.

People in ice age Britain, therefore, were hunter-gatherers living in much the same way as Greenland Eskimos or Australian Aborigines did until very recently. The all-important food quest would have concentrated on meat, obtained by killing animals or scavenging from carcasses, and collecting such food as berries, nuts, leaves, birds' eggs and snails. Adding to the difficulties of an uncertain food supply, large and powerful carnivores, such as lions and bears, lived in ice age Britain, as did formidable plant-eaters, including elephants and rhinoceroses which occasionally would have threatened life and limb.

This book is concerned with the fascinating interrelationships of humans and wildlife in ice age Britain, as revealed principally by the rich fossil record of animals and plants preserved in the sediments of rivers, lakes and in caves. It is not a book about artefacts, except in the sense that they provide most of the evidence for human presence in ice age Britain, and to a limited extent for the information that they provide on hunting techniques and other aspects of life in the ice age. For detailed and authoritative accounts of British palaeolithic archaeology, the reader is referred to the bibliography (chapter 10).

Active work on ice age Britain includes research into the dramatic changes of climate, the complex deposits laid down by glaciers and ice sheets which at various times covered much of Britain and northern Europe, and changes in sea levels and the resulting connections and separations of Britain from mainland Europe, in addition to research on fossils and archaeology.

It is against the rich background of ice age environmental changes that we view the biological and cultural evolution of man in the British Isles.

2
The ice age in Britain

In this book the informal term 'ice age' is used to cover the period from about 500,000 to 10,000 years ago, equivalent to the Middle and Upper Pleistocene or Quaternary period in geological terminology. While evocative, 'ice age' is a little misleading. Although during most of the last half million years the climate was indeed considerably colder than now, it was only for relatively short periods, widely separated in time, that ice sheets or glaciers spread over large areas of northern Eurasia and North America. The usual climatic conditions in Britain and north-west Europe over this timespan were cold and dry, producing treeless 'steppe-tundra' vegetation, with grasses, sedges and abundant herbs. These relatively stable conditions were interrupted by phases of intense cold with little vegetation (*polar desert*) culminating in widespread glaciation, but also by temperate phases (interglacials and interstadials) which resulted in the spread of forests northward and westward into the British Isles. It should be stressed that both extremes of climate took up only a small fraction of the ice age, and that the old idea that interglacials alternated with glaciation phases of approximately equal length is incorrect.

The term *interglacial* is used to describe major warm phases, each of which appears to have lasted 10,000 to 15,000 years, whereas *interstadial* is used for warm phases of lesser duration and/or intensity. The distinction, however, is merely one of degree; a warm phase recognised as an interstadial in northern Europe may have had all the characteristics of an interglacial further south.

We are now living within a major warm phase, which has so far lasted about 10,000 years (the warm-up began as early as 13,000 years ago.) It compares closely with previous interglacial periods in such features as the natural development of the vegetation, and providing that human activities, such as burning of fossil fuels, do not fundamentally affect the climate, it seems likely that colder conditions will return within a few thousand years from now. In other words the ice age which conventionally ended 10,000 years ago, is in reality probably not yet finished.

Sediments and stratigraphy

The ice age is represented in Britain and Europe by a variety of sediments deposited in rivers, lakes, the sea and in caves by melting glaciers, wind and water. To a large extent the types of sediment deposited are controlled by climatic conditions. Phases of extreme cold are represented by till (boulder clay), formed from the debris of melting glaciers and ice sheets, together with outwash sands and gravels laid down by the escaping meltwaters. Winds carried silts and sand away from the ice sheets some hundreds of kilometres, to be deposited as thick sheets of loess and coversands. In contrast the deposition of sediments such as organic muds (plant detritus) in rivers and lakes is generally characteristic of interglacial or interstadial phases.

The deposition of sediments in caves is complex and as yet poorly understood, but most of the characteristic deposits of red-brown or yellowish silts found in caves (cave earths) originate in warm phases, whereas breccias (frost-shattered rock debris) date from colder times.

The stratigraphy of ice age deposits, that is the sequence in which the various beds were laid down, is especially difficult to interpret since long vertical sequences very rarely occur at a single site. Thus, the stratigraphy has mostly been pieced together from fragmentary sequences scattered over a wide geographical area. Heavy reliance has been placed on correlating, or matching, organic deposits, using similarities in their fossil content in sequences of otherwise unfossiliferous sediments. As for older geological periods, the principle is that, since the fauna and flora change with time, deposits can be assigned to a particular interglacial or other period on the basis of a distinctive fossil content. Interglacials were not, however, periods of uniform environmental conditions; each shows a cycle of changes in climate, vegetation and animal life. The detailed correlation of sequences of fossils (especially pollen — recording the changes in vegetation) is a powerful tool for making sense of the complex geological record.

Radiocarbon dating (see below) gives absolute dates in years for suitable fossil organic materials. At present it can date back to a maximum of about halfway through the Last Cold Stage (Devensian). Although progress is being made with other absolute dating techniques which extend further back, no reliable framework of dates in absolute years is available for most of the ice age.

Table 1 shows the stratigraphical stages of the ice age (Middle and Upper Pleistocene) as currently recognised in Britain, and their approximate correlations with those of mainland Europe and North America. Many workers are, however, dissatisfied

British Names	*Netherlands/Germany*	*North America*	*Years ago (at beginning of stage)*
Flandrian/ Postglacial	Flandrian/ Postglacial	Holocene/ Postglacial	10,000
DEVENSIAN/ LAST COLD STAGE	WEICHSELIAN/ LAST COLD STAGE	WISCONSINAN/ LAST COLD STAGE	110,000 approx.
Ipswichian/Last Interglacial	Eemian/Last Interglacial	Sangamonian/ Last Interglacial	120,000 approx.
WOLSTONIAN	SAALIAN	?	
Hoxnian	Holsteinian	?	? 200-250,000
ANGLIAN	ELSTERIAN	?	
Cromerian	Cromerian	?	? 350-500,000

Table 1. Stages of the ice age (Middle and Upper Pleistocene). Cold stages are indicated by capital letters; interglacials by small letters.

with this scheme and postulate additional stages, especially an extra interglacial between the Hoxnian and Ipswichian. Important support for this is shown in analyses of cores from the deep ocean floors, which appear to show more fluctuations in climate than are recorded on land. Although there may have been additional interglacials within the last half million years, so far none of these has been convincingly demonstrated. Further advances in absolute dating techniques may eventually resolve this important question.

Within each cold stage at least one major advance of the ice sheets or glaciers occurred. During the Anglian glaciation ice advanced as far south as Finchley, North London, across to the Severn estuary. In the succeeding Wolstonian glaciation much of Britain was again covered by ice, although less extensively, but the precise limits are not known. The last major glaciation of the British Isles occurred in the later Devensian stage, about 18,000 to 15,000 years ago, when most of central and southern Britain remained ice-free.

Absolute dating

Most absolute dating methods are based on the radioactive decay of certain isotopes (forms of an element with different atomic weights) which are unstable. Decay proceeds at a known rate for each isotope independently of external factors such as temperature, pressure and light. Dates in years are obtained by measuring either the fraction of the original 'parent' isotope remaining, or the percentage of new 'daughter' isotopes pro-

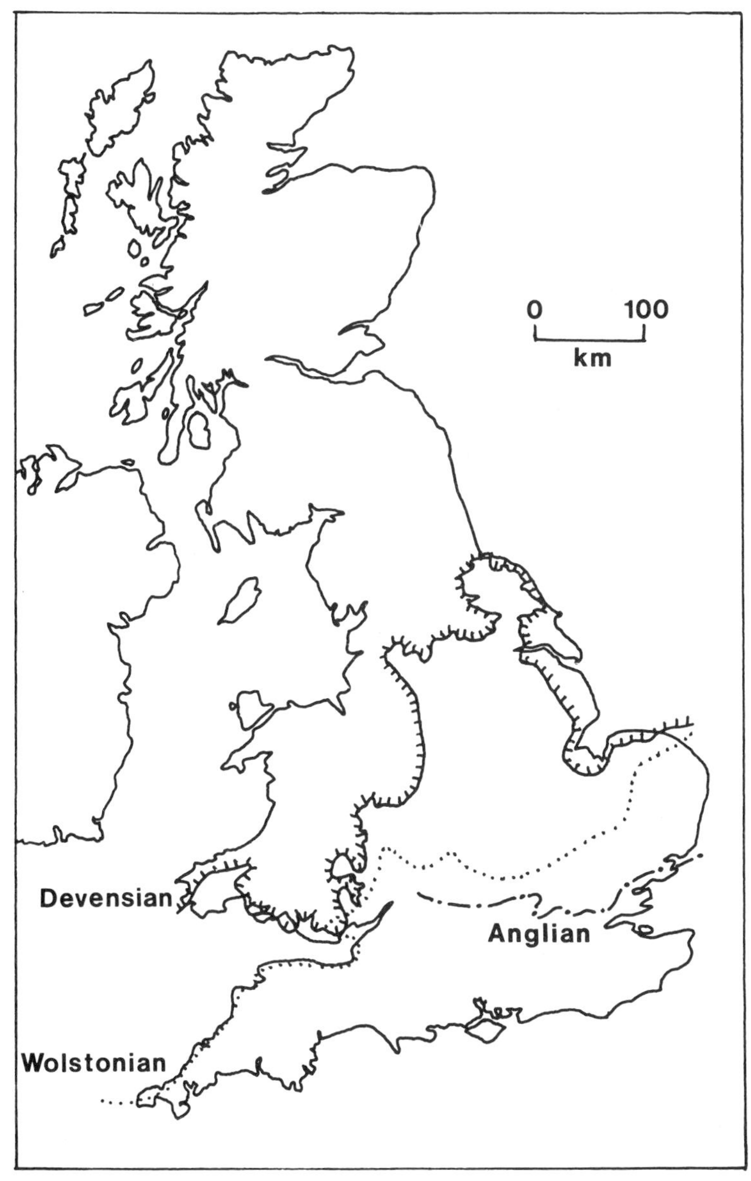

1. Limits of glaciation in Britain during the ice age. The Anglian is the oldest period of glaciation, the Devensian the youngest.

duced. The three main methods in current use will be considered here: radiocarbon, applicable at present to about 50,000 years ago; uranium-series dating of cave stalagmites, extending back about 300,000 years; and thermoluminescence dating, with a similar range.

The radiocarbon technique uses the fact that carbon-14 atoms, continuously produced in the earth's atmosphere, are absorbed by plants during photosynthesis and distributed throughout the tissues of all living organisms, which are therefore slightly radioactive. After death, carbon-14 is no longer absorbed and the radioactivity of the remains slowly decreases. Measurement of the remnant radioactivity, or the proportion of radioactive to stable carbon isotopes, of a sample of fossil organic material can therefore give an estimate of the age in years.

Uranium-series dating measures the small amounts of uranium incorporated in crystalline calcium carbonate of stalagmites forming in limestone caves. Since the daughter isotopes, produced by radioactive decay of the parent unstable uranium isotopes, were not incorporated into the stalagmite when it formed, measuring the proportions of parent to daughter isotopes gives an estimate of actual age.

The thermoluminescence method involves measuring the output of light on heating a sample of sediment or reheating a burnt flint. A substance exposed over a long period to bombardment by particles and radiation originating from small quantities of radioactive minerals, occurring widely in the ground, accumulates unstable electrons. Controlled heating of the sample results in the freeing of these electrons, accompanied by the emission of a light signal, which is measured. The rate at which the sample is bombarded with particles is assessed using detectors at the sample site, and from these measurements a date is calculated.

Radiocarbon dating is now a well established method and the results obtained are generally reliable. The other methods are still uncertain, and for the time being their results should be interpreted with caution.

Climatic change

Marked climatic fluctuations are an outstanding feature of the ice age, although the prevailing climate was colder and drier than today.

The presence of ice sheets over large areas of the British Isles shows dramatically that there were phases of extremely cold climate. Evidence is also given by fossil ice-wedge casts, commonly seen in section in gravel pits and similar places as infilled deep cracks penetrating down from a former land surface. These resulted from the cracking of permanently frozen ground (permafrost) and occur today in arctic areas where mean annual temperatures are -16 to -18 F (-6 to -7 C).

In marked contrast, plant fossils in interglacial deposits indicate temperatures similar to those of the present day (see chapter 3). Trees represented include such species as oak, elm and lime. Moreover, records of plants like water chestnut and animals such as European pond tortoise, well north of their present ranges, shows that during interglacials temperatures were a little higher than now. The Ipswichian/Last Interglacial was especially warm, resulting in the spread to England of southern species of land and freshwater mollusc and beetle, for example a dung beetle now restricted to southern Europe.

Differences in assemblages of fossil plants suggest that the Hoxnian Interglacial may have had a more oceanic climate (cool summers and mild winters) whereas the second half of the Ipswichian Interglacial was unusually continental — drier, with more extreme winter and summer temperatures.

By comparing the climatic tolerance of present-day beetles,

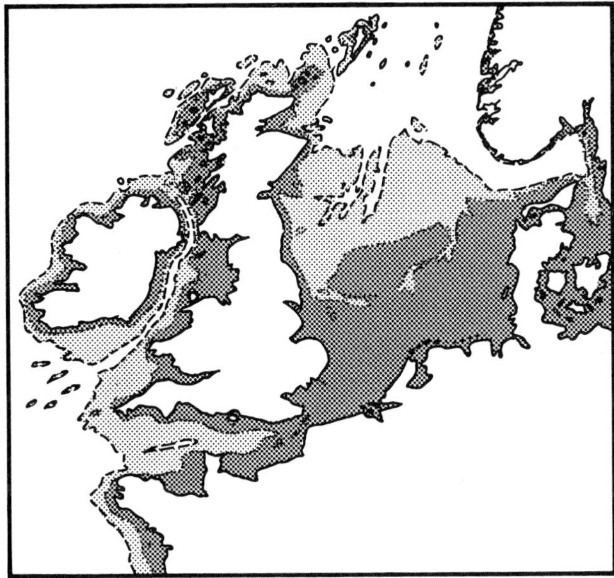

2. Contours around the British Isles show the approximate effects of lowered sea levels during the coldest phases of the ice age. Depths of 100 metres (328 feet) below sea level are indicated by light stipple; depths of 50 metres (164 feet) by heavy stipple.

distributions of fossil beetles (dated by radiocarbon) have been used to reconstruct climatic changes for the Devensian/Last Cold Stage. Some of the changes recorded using beetles are not seen in the fossil plant record, in particular an apparently rapid phase of warming about 43,000 years ago (Upton Warren Interstadial), in the middle of the Devensian (see chapter 3).

Studies of fossil plankton from deep-sea sediments show that during cold phases the 'polar front' — the boundary between warmer and polar waters in the North Atlantic — was pushed as far south as Spain and Portugal, holding the British Isles in an icy grip. This is in marked contrast to the interglacials, when the climate of Britain was warmer than expected for their latitudinal position because of the Gulf Stream Drift. The contrast between cold and warm stages is greatest for the British Isles and North Atlantic seaboard, decreasing southwards and eastwards into continental Europe.

Sea level changes and connection of Britain to continental Europe

The occurence on land of sediments containing fossils characteristic of brackish or marine conditions suggests higher interglacial sea levels than today. In some cases, however, vertical land movements may have to be taken into account, which complicates the interpretation of past geography.

During cold phases vast volumes of the world's water were locked up in ice sheets, resulting in spectacular drops in global sea levels — up to 100 metres (328 feet) or more. Even a modest drop of 50 metres (164 feet) was enough to connect Britain to Europe across the southern North Sea basin, while a 100 metre drop would have resulted in an even broader connection, perhaps also joining Britain to the north of Ireland via a narrow isthmus from south-west Scotland.

Although interglacial sea levels were as high or higher than today, Britain apparently remained connected to Europe throughout the Cromerian and Hoxnian interglacials as well as in the cold stages. The Straits of Dover were probably first flooded during the middle part of the Ipswichian/Last Interglacial, reverted to dry land throughout the Devensian/Last Cold Stage and were again flooded about 8500 years ago in the Flandrian/Postglacial. During most of the ice age Britain was a peninsula of Europe, allowing free immigration of plants, animals and humans. The absence from Ireland of many species of ice age mammals and of man until after 9000 years ago, however, strongly suggests that there was no land connection with Britain.

3
Flora and fauna

Most evidence for the changing environments of ice age Britain comes from the fossil remains of plants and animals. Their modes of occurrence, the methods used in interpretation and some of the more important conclusions reached are outlined in this chapter.

Vegetation

All flowering plants, especially those pollinated by wind, produce enormous numbers of microscopic pollen grains, which become incorporated in the accumulating deposits of lakes, rivers and peat bogs. Samples of fossil pollen are taken at intervals — usually 5 or 10 cm (2 or 4 inches) — through a sediment profile, either exposed in section as in a cliff or quarry or in a core from a borehole, and the pollen grains extracted, identified and counted using a microscope. Studies of modern 'pollen rain' show that the percentages of various pollen types in a sample generally reflect the proportions of plant species in the present-day vegetation. Percentages of fossil pollen therefore can be interpreted to reconstruct vegetational conditions at various times in the past and a sequence of samples records changes in vegetation with time.

Plant macrofossils, that is the larger remains — mostly fruits and seeds, obtained by sieving sediment — give direct evidence of which plants grew in the vicinity of the site. In many cases this allows more precise identification than is possible from pollen grains. Pollen gives a broad regional picture of the vegetation cover, so the two sources of information are complementary. Much of our knowledge of fossil plants and vegetational changes for the interglacial stages is due to work done at Cambridge by Professor R. G. West and others.

The pollen diagrams for each interglacial record a generally similar pattern of vegetational development, allowing subdivision into four 'pollen zones', designated I-IV (see figure 3). These vegetational changes were in response to changes in climate and to a lesser extent soil composition. At the beginning of an interglacial the climate warmed, resulting in the migration of birch and pine into Britain from the south and east in continental Europe (pollen zone I). Then followed a long period (several thousand years) of a climate as warm or warmer than now,

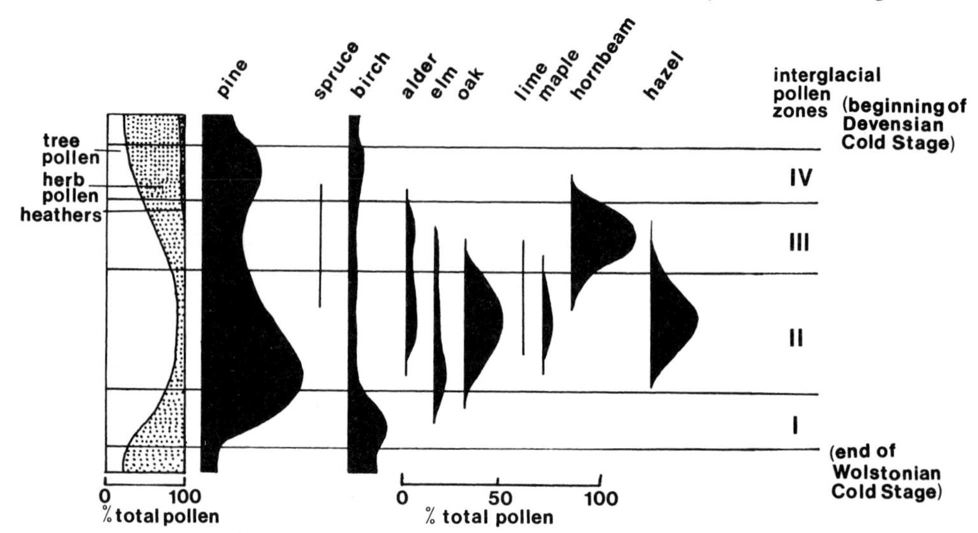

3. Pollen diagram recording changes in vegetation through an interglacial. (The example given is for the Ipswichian/Last Interglacial.) The curves represent percentages for each species of total fossil tree-pollen grains. The small block on the left records the changing dominance of forest in relation to open ground, in the form of tree pollen versus herb pollen percentages.

differing to some extent from one interglacial to another and very probably including some minor fluctuations. The regional vegetation cover was temperate broadleaf 'mixed oak forest' (zones II and III). After this phase climatic deterioration set in and interglacial conditions gradually gave way to those of the succeeding cold stage, with the return of birch and conifer forest (zone IV). The sequence of vegetational changes as recorded in the pollen diagrams appears to be characteristic for each interglacial.

Fossil plants are known from each cold stage, especially the early and late phases immediately following or preceding an interglacial stage. The complex history of vegetational development in cold stages, however, is only known in any detail for about the last 40,000 years of the Last Cold Stage or Devensian (approximately 50,000 to 10,000 years ago). The predominant type of vegetation of the colder periods both in Britain and now temperate areas of north-west Europe has no exact modern analogue, but appears to have resembled both modern arctic tundras and the steppes of central Eurasia. This 'steppe-tundra' vegetation was characteristically treeless with grasses, sedges, arctic alpines, wormwood, *Artemisia* and other herbs as major components. Shrubs included dwarf birch, arctic willow and juniper.

During many of the milder climatic phases which punctuated the prevailing cold, tree birch and pine were able to colonise parts of Britain and form extensive woodlands or forests. During the Chelford Interstadial (named after a locality in Cheshire), in the early part of the Devensian, spruce was also present. The forest was apparently very similar to that found in northern Finland at the present day.

Animals

Under favourable conditions a variety of fossil animal remains can be preserved. The most important of these are beetles, freshwater and terrestrial molluscs and vertebrates (animals with backbones: fishes, amphibians, reptiles, birds and mammals).

Beetles have a tough external skeleton of a substance called chitin, resistant to most forms of chemical attack, including the humic acids present in peats. The fossils occur mostly as remains washed or blown in from the immediate surroundings. Studies on the Devensian Cold Stage by Dr G. R. Coope and his co-workers have shown for example the presence of beetles now confined to the Arctic or central Asia, while the few interglacial faunas so far studied include species now found much further south. Of particular interest is the presence of temperate species of beetle in faunas from particular phases within the Devensian. In particular the Upton Warren Interstadial, radiocarbon dated to about 43,000 years ago, implies summer temperatures at least as warm as today, whereas fossil plant evidence indicates only a tundra-like vegetation consistent with arctic climatic conditions. Perhaps the warming of the climate was both rapid and short-lived, not allowing sufficient time for trees to migrate into Britain, although beetles were able to respond more quickly.

The shells of molluscs, including snails, living on land and bivalve 'mussels' in fresh water, are common fossils from ice age deposits. They represent animals which lived and died in the water, together with those found on marsh vegetation or washed in from dry-land habitats. Series of samples taken through a sediment profile allow the construction of a 'mollusc diagram', as is done with pollen and beetle samples. Molluscs can give a great deal of useful ecological information about past environments, such as whether water was running or still and bordered by marsh, dry grassland or woodland. Fossil records of certain species provide important additional data on interglacial climates (see chapter 2).

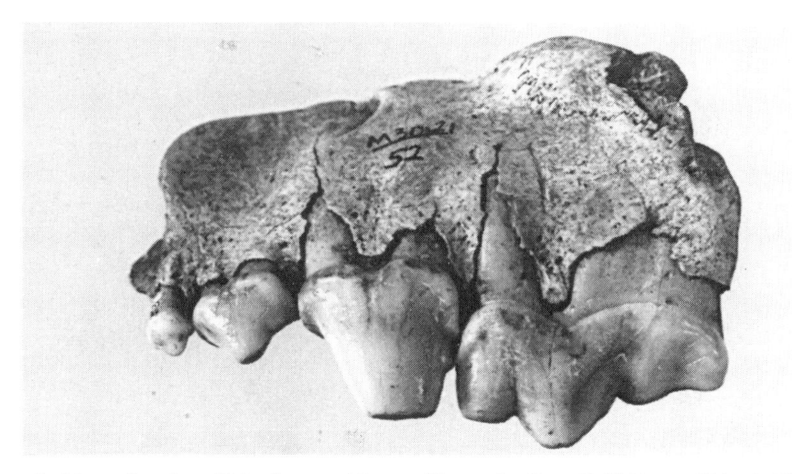

4. Upper jaw (maxilla) of spotted hyena (Devensian/Last Cold Stage, Picken's Hole, Somerset). Note the powerful conical premolar teeth used for crushing bones and to the rear the shearing carnassial tooth for cutting through flesh. In common with the many bones and teeth of a variety of animals from this small cave site (see figure 5), this jaw has been itself extensively chewed and damaged by other hyenas. Length of specimen 94 mm (3¾ inches) (Photograph: author.)

5. Molar teeth of baby mammoths (Devensian/Last Cold Stage, Picken's Hole, Somerset), brought into the cave by spotted hyenas. Hyenas were not able to carry the jaws or teeth of adult mammoths. (Photograph: author.)

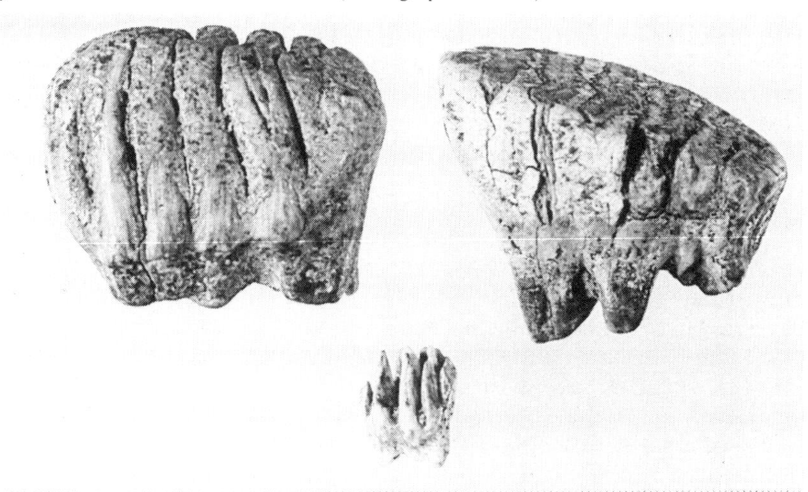

6. Skull and lower jaw of woolly rhinoceros from river deposits (Devensian/Last Cold Stage, Coston, Norfolk). (Photograph: author.)

Remains of vertebrates are relatively common in ice age deposits. The form of each part of the skeleton is generally diagnostic and allows identification to genus or species, even when only part of a bone or tooth is preserved. The teeth in mammals are especially useful for identification purposes. Birds show less range in skeletal form, and are consequently difficult to work with. Moreover, bird remains are generally rare and there are numerous present-day species which could conceivably be present in a fossil assemblage. The rarity of fossil birds, except in some cave deposits, is no doubt because of the fragility of their remains. Tiny vertebrate remains such as fish scales, snake vertebrae and shrew jaws are found only in fine-grained sediments (deposited in a lake, river or cave) with a high content of lime (calcium carbonate), often shown by the presence of numerous mollusc shells in the same deposit. Remains of large animals, especially the teeth and larger bones of elephants, are commonly found in coarse sands and gravels from which smaller vertebrate fossils have been leached away by acidic waters in the ground. Collagen, a protein found in bone, survives in many of the fossils, especially the more recent ones, and it is this substance which can be dated by the radiocarbon method (chapter 2).

There is generally no movement of the coarse sediments (sand and gravel) from the margins of a lake to the centre, so sediments there are usually fine-grained silts, clays and 'organic muds' (deposits consisting largely of finely divided plant remains). Vertebrate fossils in such sediments are mostly from animals such as fishes and amphibians (frogs, toads and newts) which lived and died in the lake. Occasionally a large mammal skeleton, such as a mammoth or a giant deer, is found in the deposits of a former lake. In such instances the animals appear to have perished in a natural trap, either becoming mired in soft sediments at the lake margin, or falling through thin ice in winter. Archaeological material is sometimes preserved in lake deposits, representing accumulated rubbish from lakeside hunting encampments. This appears to have been the situation at Hoxne (chapter 5).

Vertebrate remains preserved in river sediments include those of animals which lived and died close to where the sediments were deposited. In addition, because rivers actively erode their banks, the sediments are also generally rich in bones and teeth of animals which lived on dry land. Accumulations of complete or partial skeletons may have originated from corpses carried for some distance. Most of the mammal remains from river sediments, however, appear to have been incorporated as the river eroded its banks, sweeping in bones and teeth of various ages and states of preservation from the land surface. Such fossil assemblages therefore include a wide variety of species. Bones commonly bear the tooth marks of spotted hyena or wolf. Much of the small-mammal material probably originated from pellets discarded by roosting owls or other birds of prey, or from the droppings of smaller carnivores, such as foxes.

The richest sources of fossil vertebrate remains are undoubtedly caves, where bones and teeth have become concentrated within the lime-rich deposits. Most of the larger mammal remains in British 'bone caves' were evidently brought in by spotted hyenas or wolves, as is shown by characteristic damage inflicted by the

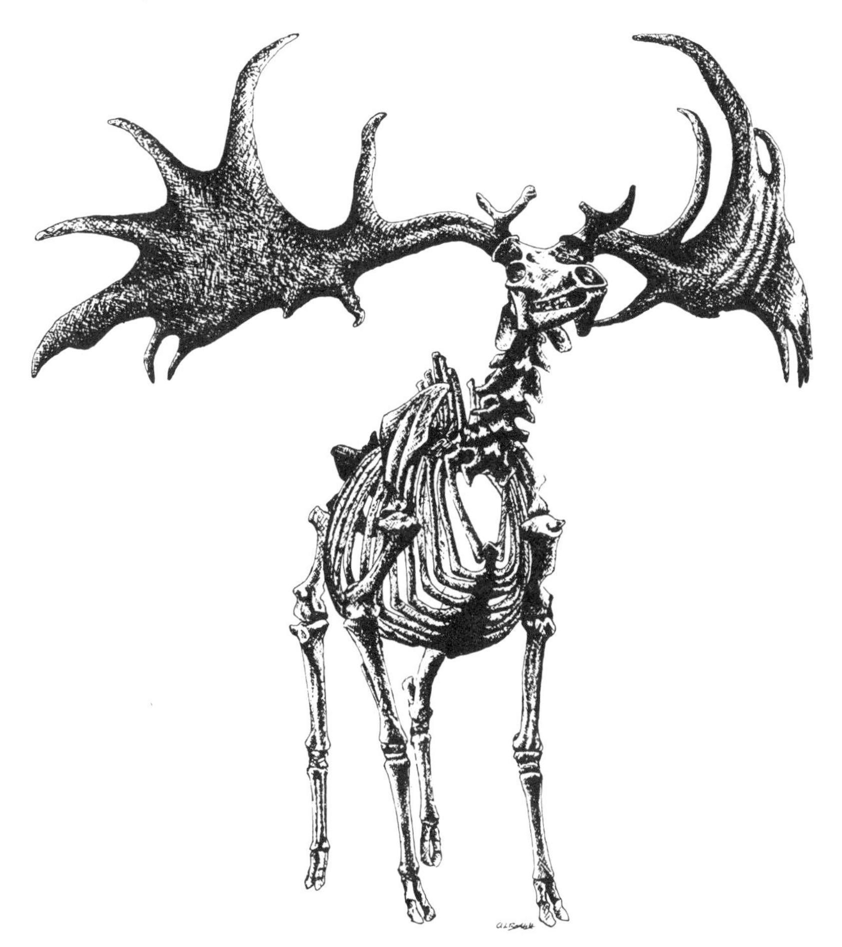

7. Skeleton of giant deer *Megaloceros giganteus* from Limerick, Ireland (Late Devensian/Last Cold Stage). (Drawing of exhibit in Sedgwick Museum, Cambridge, by Anna Bartlett.)

teeth of these predators and scavengers. Material from a former hyena den is, with experience, unmistakable, revealing a high degree of bone fragmentation, preferential preservation of the teeth and more robust bones, and polish seen on fragments corroded by the hyena's powerful stomach acids. Hyena den assemblages invariably also contain a high percentage of the remains of the hyenas themselves, which have been treated just as disrespectfully as the other material. Humans have also been reponsible for accumulating animal bones and teeth in caves, mainly as food refuse. However, unless humans are the only predators represented in a cave deposit and there is little or no carnivore tooth damage on the material, the mere association of artefacts with animal bones in the same bed cannot be taken as proof that the animals were killed and eaten by man. Until fairly recently it was commonly assumed that all such associations provided direct evidence of human diet and hunting in the palaeolithic period.

The vertebrates recorded from cold stages are generally very different to those from interglacials. Typical 'cold' faunas comprise a mixture of species today confined to arctic regions (for example, lemming, reindeer, arctic fox) or the steppe grasslands of eastern Europe and central Asia (for example saiga antelope, ground squirrel) and extinct animals (for example woolly rhinoceros, mammoth, giant deer). In addition a few species occurred which are still found today in temperate Europe and were present also in interglacials (for example water vole, red deer). Even more remarkable are the numerous records of lion and spotted hyena from the middle part of the Devensian/Last Cold Stage, about 30-45,000 years ago. Spotted hyena is now confined to Africa south of the Sahara and lion, even in historical times, was absent from Europe with the apparent exception of a small area of the south-east Balkans. This curious combination of animals reflects conditions of climate and vegetation which have no exact present-day equivalent.

In contrast, interglacial faunas consist largely of animals still found in Britain and adjacent parts of continental Europe, such as fallow deer and wild boar, both characteristic of temperate broadleaf (deciduous) woodland. Extinct forms, progressively less numerous in younger deposits, include straight-tusked elephant and rhinoceros of the genus *Dicerorhinus*. The presence of a few species now found only well to the south of the British Isles (for example hippopotamus, pond tortoise) is consistent with other evidence that interglacial climates were generally warmer than at present.

4
The earliest immigrants

The Cromerian Interglacial

Along much of the coastline of Norfolk and part of Suffolk, in the cliffs and on the foreshore, is found a series of marine and freshwater silts, sands, gravels and other deposits known as the Cromer Forest Bed Formation. The Cromer Forest Bed is especially well known for the fossil mammal remains which occur abundantly. They were laid down over a long period covering several major oscillations of climate, but we are concerned only with the latest and best known of the phases, the Cromerian Interglacial, probably dating back to between 350,000 and half a million years ago.

At West Runton, near Cromer, Norfolk, a dark peaty deposit — the West Runton Freshwater Bed — was laid down by a slow-flowing river during the first half of the Cromerian Interglacial. Although the deposits have yielded abundant fossil remains of both plants and animals, so far no trace of humans has been found. Apparently man had not yet arrived in Britain. This absence in itself is of great interest: it enables us to see what the wildlife of southern lowland Britain was like before any human interference.

At West Runton the fossil plant remains — fruits, seeds and microscopic pollen grains — studied by Professor R. G. West (University of Cambridge) record the development, in the middle of the interglacial (pollen zone II) under a temperate climate possibly a little warmer than now, of a regional mixed oak forest. This forest, which would have clothed much of Britain, had extended across the dry bed of what is now the southern North Sea basin from continental Europe. Its constituents included oak, elm, spruce, pine, birch, alder, lime and hazel. The former river at West Runton supported many aquatic plants, with reed swamp and fen fringing the shore. On nearby drier ground grew areas of open vegetation with grasses, sedges and a rich herb flora. Fossil shells show that freshwater mussels and other aquatic molluscs lived in the river, while numerous snails and slugs were present along the river bank.

The West Runton Freshwater Bed has yielded one of the richest assemblages of fossil vertebrate remains known anywhere in Britain. Fishes recorded include rudd, roach, tench, pike, perch and three-spined stickleback — all typical of a lowland

8. Reconstructed scene from the middle of the Cromerian Interglacial, based on evidence from West Runton, Norfolk. The scene is of a river flowing through temperate forest: extinct bear on far left; spotted hyenas feeding on the carcass of a wild boar, centre

river in present-day Britain. Also represented are frogs and toads, grass snake, water birds and mammals of waterside habitats including beaver, extinct beaver *Trogontherium cuvieri,* the extinct water vole *Mimomys savini,* otter, and the remarkable Russian desman, which today survives only in the Ukraine.

Open herb vegetation on the river flood plain probably supported several species of grassland voles. As would be expected from the fossil plant evidence, much of the mammal fauna is consistent with the presence of woodland habitats. Such species include woodmouse, bank vole, fallow and roe deer, wild boar and an extinct rhinoceros *Dicerorhinus etruscus.* That the forest was not generally dense, but contained areas of open vegetation, is suggested by the presence of horse and three

foreground; extinct elk behind; stag and hind of extinct giant deer right background. There is no evidence for the presence of humans. (Norfolk Museums Service, painting by N. Arber.)

extinct species of deer in which the males carried enormous outspread antlers. Other large herbivores included red deer and an extinct bison.

Such a rich fauna of herbivores not unexpectedly provided food for a wide range of carnivores, including wolf, spotted hyena, a large cat, marten and weasel and the omnivorous extinct bear *Ursus deningeri*. Other species known from the Cromerian Interglacial but not so far recorded from the West Runton Freshwater Bed include macaque monkey, hippopotamus, straight-tusked elephant and sabre-tooth cat.

Over the years many claims have been made for the presence of humans in Britain in the early Pleistocene, one to two million years ago, but the supposed 'artefacts' on which these claims were based have failed to convince present-day archaeologists.

Until 1987, the earliest plausible evidence for human presence in Britain came from a remarkable site near Westbury-sub-Mendip in Somerset. This evidence comprises a few chipped flints interpreted by several archaeologists as of human workmanship but by others as of natural origin.

At Westbury quarrying operations in the Carboniferous limestone revealed a complex series of deposits spectacularly rich

9. Quarry face at Westbury-sub-Mendip, Somerset, as seen in 1974. (Photograph: M. J. Bishop.)

10. Section exposed in the quarry face at Westbury-sub-Mendip, Somerset. The possible artefacts and most of the mammal remains are from the 'Calcareous Member' of beds.

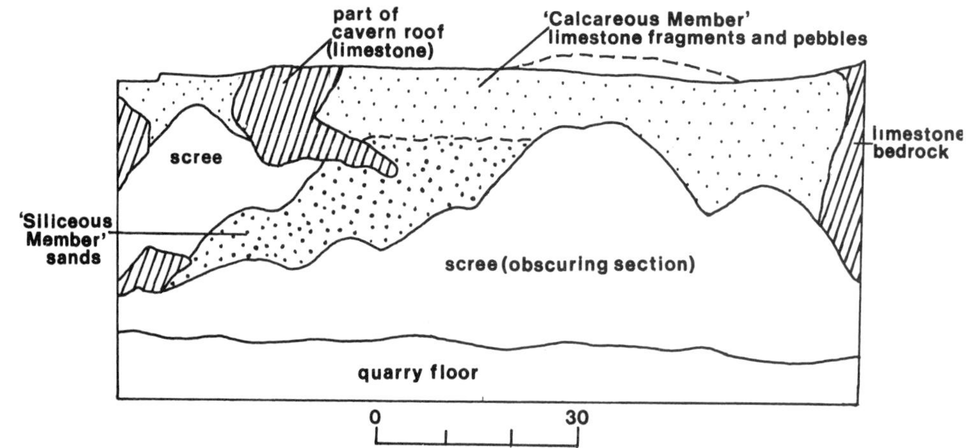

in fossil vertebrate remains, occupying a former cave system. The site was originally studied by M. J. Bishop and subsequently by a team from the British Museum (Natural History) under the direction of Dr P. Andrews. Because carnivores commonly make their dens in caves and sometimes also die there, they are usually well represented in the fossil remains from cave deposits. At Westbury the extinct bear *Ursus deningeri* is the commonest large mammal, while other carnivores include sabre-tooth cat and two species so far unique in Britain, an extinct dhole (a kind of wild dog) and an extinct leopard. The vast numbers of small mammal remains represent several species of shrew, moles, bats and many species of rodent. The presence of pond tortoise from part of the sequence clearly indicates interglacial conditions. On the other hand, at the top of the sequence, arctic lemming and other species are consistent with a cold, continental climate, suggesting the end of an interglacial or the beginning of the next cold phase.

There are many similarities between the mammals recorded from Westbury and those from West Runton, with an important exception that at Westbury the more advanced extinct water vole *Arvicola cantiana* replaces the ancestral form *Mimomys savini,* found at West Runton. This strongly suggests that Westbury is the younger site. Many workers believe that Westbury dates from a hitherto unrecognised interglacial later than the Cromerian, but pre-dating the Hoxnian. The view taken here, however, is that Westbury is most probably of late Cromerian age, especially as *A. cantiana* also occurs at Ostend, Norfolk, in deposits of the Cromer Forest Bed assigned on the basis of their pollen content to zone IV of the Cromerian Interglacial. Unfortunately the Westbury sediments do not contain good fossil pollen, and any assessment of their age relies solely on comparison of vertebrate faunas with those of other sites.

Undoubted testimony that humans had reached Britain at about the same time that the Westbury sequence was deposited is provided by the finds from Boxgrove, near Chichester, West Sussex, excavated under the direction of M. Roberts (Institute of Archaeology, London). At this remarkable site artefacts occur widely over the surface and within the top part of the Slindon Sands. These sands were deposited on top of beach deposits, banked up against a 'fossil' chalk cliff, now entirely buried by later deposits. This cliff and related deposits, now up to 41 metres (135 feet) above sea level, run approximately east-west for tens of kilometres along the West Sussex coast. The environment at Boxgrove is envisaged as a broad, flat, marshy coastal plain to the

11. Excavation in progress at Boxgrove, West Sussex, a site which has produced clear evidence of humans in association with animal remains characteristic of the Cromerian Interglacial. (Photograph: Institute of Archaeology, London.)

south of a degraded cliff line.

The artefacts comprise several fine flint handaxes of Acheulian type (named after a locality at St Acheul, France) and numerous flakes. It is thought that tools were manufactured on the surface of the sands, using as raw material flint nodules collected from a collapsed part of the old cliff nearby.

Bones and teeth of a wide variety of animals were found scattered on and within the Slindon Sands. Mammal remains, studied by A. P. Currant (British Museum, Natural History) and S. Parfitt (Institute of Archaeology), include an extinct rhinoceros *Dicerorhinus etruscus,* and various voles and shrews known from both West Runton and Westbury but not from geologically younger deposits. The presence of an advanced water vole *Arvicola cantiana,* however, suggests that the Boxgrove material is, like Westbury, rather younger than that of West Runton. Pollen analyses suggest the presence of mixed conifer/deciduous woodland consistent with the later part of an interglacial. Again the suggestion has been made that these deposits represent a 'new' interglacial between the Cromerian and the Hoxnian, but they could equally well date from the later part of the Cromerian Interglacial.

The Anglian Cold Stage

In this period, ice sheets extended as far south as London and across to the Severn estuary, further than in any other glacial episode of the ice age. Fossil ice-wedge casts record the occurrence of permafrost (permanently frozen ground) at various times during this stage. As with other cold stages, however, the coldest phases accompanied by glaciation and/or permafrost probably occupied only a fraction of the entire Anglian stage. Such evidence as we have indicates that the fauna and flora of much of the Anglian, accompanying less severe climates, were generally similar to those of later cold periods, with predominant grassy herbaceous vegetation supporting such animals as reindeer, Norway lemming, horse, woolly mammoth and woolly rhinoceros.

There is no definite evidence for the presence of humans in Britain at any time during this stage, although water-worn artefacts, presumably derived from late Cromerian deposits, are known from some deposits thought to date from this stage.

12. Handaxe of Acheulian type, Boxgrove, West Sussex. (Photograph: Institute of Archaeology, London.)

13. Lower jaw (mandible) of beaver, Boxgrove, West Sussex. (Photograph: Institute of Archaeology, London.)

14. Lower jaws (mandibles) of red deer from the main assemblage of bones and artefacts ('Lower Industry') at Hoxne, Suffolk. The scale is 10 cm (3.9 inches) long. (Photograph: J. J. Wymer.)

15. Skull and antler bases of red deer from the main assemblage of bones and artefacts ('Lower Industry') at Hoxne, Suffolk. The scale is 10 cm (3.9 inches) long. (Photograph: J. J. Wymer.)

The Hoxnian Interglacial

In marked contrast to the general paucity of evidence for man in earlier periods, many deposits assigned to the Hoxnian Interglacial, from about 200-300,000 years ago, contain abundant artefacts. In addition, scarce human skeletal remains have also been found.

At Hoxne (pronounced 'Hoxen'), Suffolk, the locality after which the interglacial is named, a large hollow in the deposits of till, or boulder clay, left by the melting ice of the previous Anglian Cold Stage, filled with water, giving rise to a lake. Over thousands of years pollen was incorporated in the clays and silts accumulating in lake bed and thus recorded the changes from a subarctic tundra-like vegetation at the end of the Anglian, through a succession of forest types during rather more than half of the temperate Hoxnian Interglacial.

As long ago as the late eighteenth century John Frere found flint handaxes in the Hoxne brickearths and with remarkable insight observed that they were 'fabricated and used by a people who had not the use of metals . . . belonging to a very remote period indeed, even beyond that of the present world'.

Excavations from 1974 to 1978 directed by J. J. Wymer as part of a project sponsored by Professor Singer of the University of Chicago, revealed several beds containing artefacts in the sands, gravels and silts overlying the ancient lake deposits. The main assemblage of stones, flint handaxes of Acheulian type and flakes — the 'Lower Industry' — together with animal bones and teeth, occurs in a bed of silts just above the top of the lake deposits. The presence amongst the fossil material of such animals as macaque monkey and fallow deer, taken with the position of the bed in relation to the lake deposits, strongly suggests a date within the later part of the Hoxnian Interglacial (pollen zone IV?). In the absence of a reliable fossil pollen record, evidence for environmental conditions comes almost entirely from the fossil vertebrates. Finds of rudd, tench and pike are consistent with the presence of a body of fresh water, which also attracted such amphibious animals as otter, an extinct water vole, beaver, extinct beaver *Trogontherium cuvieri* and Russian desman — all recorded also from the Cromerian.

Temperate woodland is indicated by the remains of fallow deer, macaque monkey, an extinct rhinoceros *Dicerorhinus sp.*

16. Reconstructed scene from the later part of the Hoxnian Interglacial based on evidence from Hoxne, Suffolk. Around the lake are shown: macaque monkeys; Russian desman (in foreground); extinct beavers *Trogontherium cuvieri* (middle distance and in water); horses; fallow deer; an extinct rhinoceros *Dicerorhinus sp.*; elephants; and handaxes and other artefacts left by humans. (Line drawing by N. Arber.)

and less definitely by roe deer. At the same time extensive areas of open vegetation are suggested by such animals as horse (the animal most abundantly represented), giant deer, lion, short-tailed vole and Norway lemming — here as at several other localities apparently indicative of open vegetational conditions, but not necessarily also of a cold climate. Other animals with less marked habitat preferences include red deer (in abundance second only to horse), bear and an elephant of unknown species.

It is uncertain whether or not the scattered and broken animal bones and teeth, which occur in association with artefacts, represent human food refuse. If they do there is a further question. Were humans at Hoxne active hunters or did they merely scavenge the leftovers from carnivore kills and/or the carcasses of animals which had died of natural causes (see chapter 8)? At present we have no clear answer.

At Hoxne itself, and at two other localities in East Anglia, the sequences of fossil pollen from lake deposits record a puzzling increase in grasses and a corresponding decrease in forest cover during the later part of pollen zone II. A widespread forest fire has been suggested as the likely cause of this phenomenon. In support of this is the observation at both Hoxne and one other locality (Marks Tey, Essex) of quantities of fine-grained charcoal in the sediments in precisely the correct level, although strangely charcoal is absent from a third locality, Barford in Norfolk. Since a limited forest fire would very probably lead to improved grazing and concentration of animals in the area, this raises the possibility that the fire may have been started deliberately by man. If so, it presumably got out of control, perhaps in an exceptionally dry summer, and burned a wider area than intended. There is at present no way of proving or disproving this idea. The only evidence for human presence at this time comprises two handaxes and some flakes from the lake deposits at Hoxne, and it has been suggested that these may have sunk into the sediments from a higher level. A lower jaw of straight-tusked elephant has also been found at Hoxne in the same layer that yielded the pollen evidence for deforestation.

At Hoxne the 'Upper Industry', which includes Acheulian handaxes, occurs at a higher level within the sediments overlying the lake deposits. Mammal remains from here include an extinct rhinoceros *Dicerorhinus sp.*, beaver and extinct beaver. They indicate forest and probably also a temperate (certainly a non-arctic) environment. Since the 'Upper Industry' is separated from the 'Lower Industry' by beds containing fossil arctic plants and evidence of permafrost conditions, the former appears to represent a phase of milder climate postdating the Hoxnian Interglacial. Work in the Netherlands and Germany has revealed the existence of such a temperate phase, known as the 'Domnitzian' Interglacial or Interstadial, and it is possible that the 'Upper Industry' dates from this little known phase.

At Clacton-on-Sea, Essex, numerous flint tools — cores and flakes of 'Clactonian' type — and fossil mammal bones have been obtained from the deposits laid down by an ancestral river Thames. Both the fossil pollen and mammal remains indicate that the deposits are of Hoxnian Interglacial age and that most of the fossils and artefacts probably date from pollen zone II, that is earlier than the main archaeological and faunal horizons at Hoxne. The animals represented are generally similar to those at Hoxne, although the Clacton deposits have additionally yielded wild boar. Many bones and teeth of straight-tusked elephant and aurochs (wild cow) have been found at Clacton, whereas at

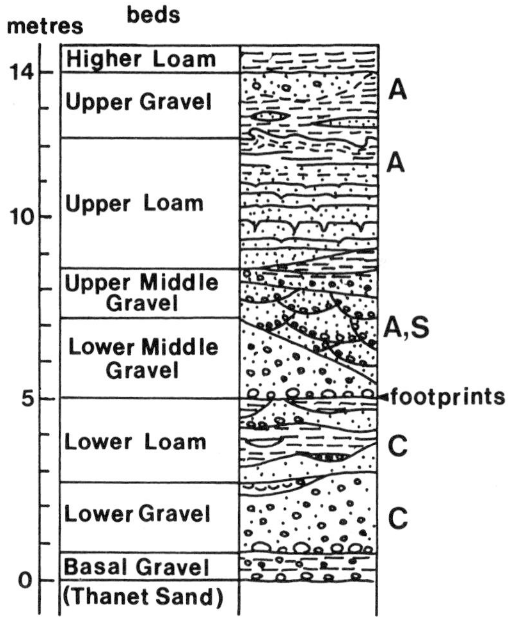

17. Section through river deposits at Swanscombe, Kent. S, horizon of human skull bones; C, Clactonian artefacts; A, Acheulian artefacts.

Hoxne elephant and bovine remains are too fragmentary for identification to species. Conversely, macaque and desman are unrecorded from Clacton. Few small-vertebrate remains of any kind have been collected from this locality.

The Clacton freshwater deposits also yielded an artefact unique in the palaeolithic of Britain — a spear point made from yew wood, shaped by working with a flint flake. This object, found by S. H. Warren early in the twentieth century, was reassessed by K. P. Oakley and others, who concluded that it could have been used only as a thrusting weapon and would have been unsuitable for throwing.

The fossil vertebrates and pollen indicate a temperate climate with broadleaf forest, comprising oak, elm, alder and other trees, with grasses and herbs growing on the river floodplain.

The best known palaeolithic site in Britain is Swanscombe in Kent, famous both for its wealth of Acheulian handaxes and other artefacts, and for the finds of human skull fragments — the earliest human remains so far known from Britain. At Swanscombe gravels, sands and silts of a high terrace of the Thames, recording a former course of that river, contain a sequence of vertebrate remains, molluscan shells and artefacts. The 'Lower

Gravel' (gravels and sands) and the 'Lower Loam' (silts and sands) at the base of the sequence contain artefacts of Clactonian type and no handaxes, a temperate fauna of freshwater and terrestrial molluscs, and a temperate mammalian fauna very much like the faunas of both Hoxne and Clacton. Mammals recorded consistent with broadleaf forest include macaque monkey, wild boar, fallow deer and straight-tusked elephant, and more open vegetation is suggested by such animals as horse, lion and giant deer. The molluscs are consistent with the presence of reedswamp and fen, with drier open ground and hazel on the river floodplain. Again the association of fossil animals and artefacts tells us a great deal about the environment of early humans but gives little information on how they exploited the available food resources.

Of particular interest with regard to the mammals at Swanscombe was the discovery and careful excavation of fossil footprints preserved in silts at the top of the Lower Loam. The footprints, which must have been made shortly before the deposition of the overlying sands and gravels, were mainly of deer, but horse, aurochs, rhinoceros and elephant are all represented. Animals probably trampled the soft silts as they came to drink at the river.

Archaeologically there was a marked change with the deposition of the 'Middle Gravels' on top of the Lower Gravels and Lower Loams. The Middle Gravels (mainly gravels and sands) contain numerous Acheulian handaxes, replacing the Clactonian artefacts of the older underlying beds. Fossil mammal remains recorded from the 'Upper Middle Gravels', although relatively few, are generally similar to those from the Lower Gravels and Lower Loam. In particular, the presence of fallow deer and straight-tusked elephant indicates the persistence of temperate woodland. As at Hoxne, Norway lemming is recorded, but no other 'arctic' species. In the author's opinion the Upper Middle Gravels are probably approximately equivalent in age to the beds containing the Hoxne 'Lower Industry', from the later part of the Hoxnian Interglacial, and the Lower Gravels and Lower Loam were probably deposited in the middle part of the interglacial, possibly pollen zone II. The Swanscombe deposits do not contain well preserved pollen and so cannot be matched directly with sequences from other localities. Recent thermoluminescence dating at Swanscombe suggests ages of about 230,000 years for the Lower Loam and about 200,000 years for the 'Upper Loam' (probably early Wolstonian) which caps the sequence.

The pieces of human cranium from Swanscombe were found in the Upper Middle Gravel in the same bed as several pointed Acheulian handaxes. The 'Swanscombe Skull', which is very similar to the more complete specimen from Steinheim, West Germany, dated to the equivalent Holsteinian Interglacial, is considered to represent an early form of *Homo sapiens,* with features suggesting it could have been ancestral to the Neanderthals and to modern man. The people of the time in western Europe appear to have been robustly built with large brow ridges (more marked in males), relatively large jaws and no chins.

The last site to be considered here is quite different to those previously discussed, as the artefacts and faunal remains occur in the sediments accumulated in a limestone cave. At Pontnewydd Cave in Clwyd, North Wales, excavations directed by S. Green (National Museum of Wales, Cardiff) revealed a sequence of deposits containing tools made by Levalloisian technique (named after a suburb of Paris where artefacts of this type were first recognised), Acheulian handaxes, mammal remains and, of special interest, fragmentary human remains, mainly teeth. Uranium-series dates on stalagmite and a thermoluminescence date on burnt flint suggest about 200,000 to 250,000 years for the 'Lower Breccia' and 'Intermediate' deposits which contain human and animal remains and artefacts. Unfortunately the deposits appear to have been disturbed by sludging when saturated with water. This process not only destroyed the original association of artefacts and faunal remains but also may have incorporated material from older beds. This possibility of mixing means that both the archaeological and faunal evidence needs to be interpreted with particular caution. The animal remains identified by A. P. Currant of the British Museum (Natural History) include the extinct interglacial rhinoceros *Dicerorhinus hemitoechus* and wood mouse, suggesting a temperate climate with woodland. The record of tundra vole, now confined to the arctic, may be based on material derived from earlier sediments laid down within a cold period. In the absence of evidence from pollen or clear indications from the fossil mammals the age of this deposit is uncertain although the absolute dates suggest that it dates either from the Hoxnian Interglacial or perhaps the 'Domnitzian' Interglacial/Interstadial. However, the occurrence of numerous artefacts made by the relatively advanced Levalloisian technique (see chapter 6) suggests that Pontnewydd could be considerably younger than Swanscombe.

18. Section of deposits in Pontnewydd Cave, Clwyd, North Wales. (Photograph: H. S. Green.)

19. Section through cave sediments at Pontnewydd. Artefacts and human remains were found in the 'Lower Breccia' and 'Intermediate' beds.

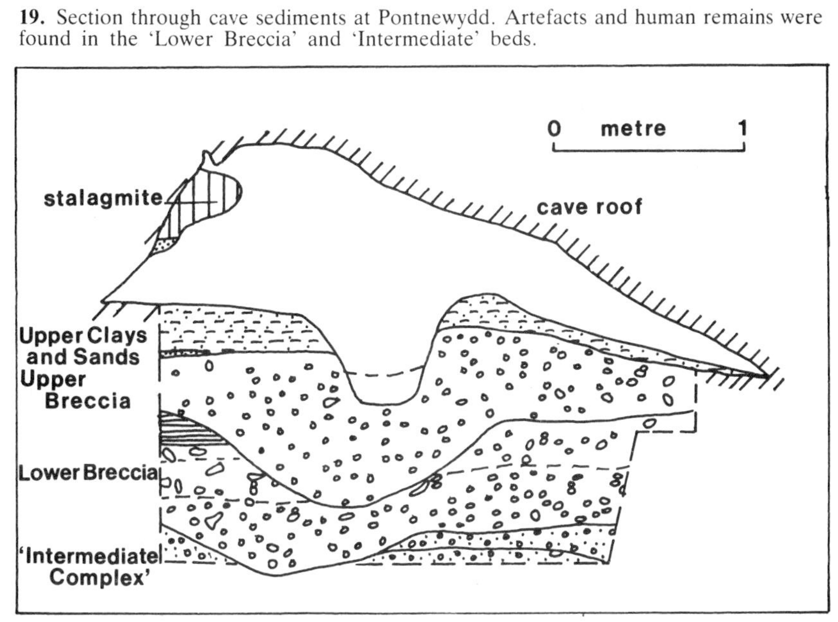

Comings and goings

During the long period that lasted perhaps 100,000 years and spanned much of the Wolstonian Cold Stage, Ipswichian/Last Interglacial and the first two thirds of the succeeding Devensian/Last Cold Stage, Britain appears to have been inhabited by humans only sparsely and intermittently. The artefacts of this period, broadly classified as middle palaeolithic (middle old stone age) and including Levalloisian and Mousterian types, are generally scarce.

Present evidence of environmental conditions during the Wolstonian Cold Stage indicates that for much of this period, as for the later and better known Devensian/Last Cold Stage, prevailing cold climates were accompanied by treeless herb vegetation and such animals as reindeer, horse, mammoth and woolly rhinoceros. Within this stage occurred a major glaciation, but there were also interstadial phases of milder climate. It is reasonable to suggest that the few artefacts from this period may all date from interstadial phases and that humans were otherwise absent, although this is controversial. In the early twentieth century at Baker's Hole, Northfleet, Kent, flint artefacts were found in deposits which are generally thought to date from the end of the Wolstonian, although they may be of Ipswichian age. The artefacts include large flakes struck from prepared cores by a characteristic technique, termed Levalloisian, and handaxes of Acheulian type.

Further south in continental Europe the situation was different: deposits covering approximately the same period are rich in artefacts, presumably reflecting warmer climates and more abundant and reliable food supplies. At La Cotte de St Brelade, Jersey, Channel Islands, excavations revealed a sequence of deposits of Wolstonian age filling a deep fissure or ravine in the granite bedrock, and underlying the pebbles of a beach dating from the Ipswichian/Last Interglacial. The deposits, consisting mainly of wind-blown loess derived from glacially deposited sediments hundreds of kilometres to the north and east, have yielded vast numbers of artefacts. Lowered sea levels at this time meant that the Channel Islands were then broadly connected to the French mainland, allowing free passage of animals and people. Of particular interest at two horizons are 'bone piles' including many bones and several complete skulls of mammoth

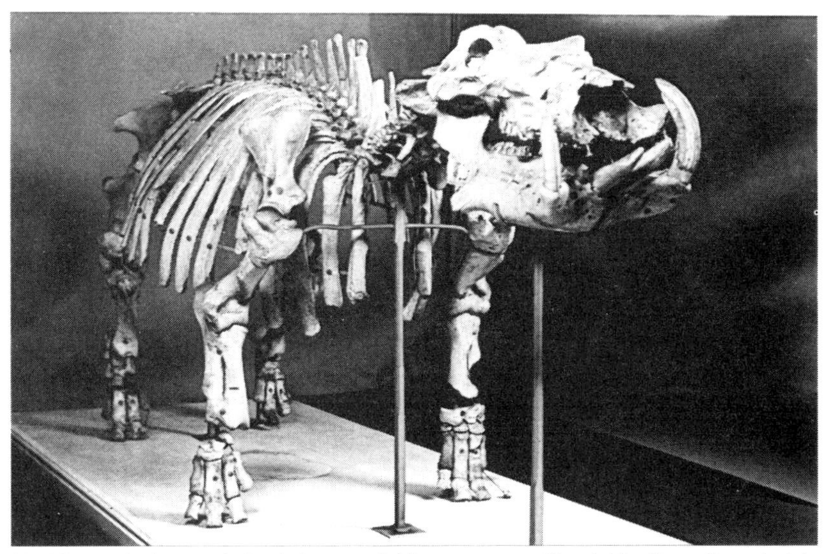

20. Mounted (composite) skeleton of hippopotamus (Ipswichian/Last Interglacial, Barrington, Cambridgeshire). (Photograph: Sedgwick Museum, Cambridge.)

and woolly rhinoceros. Study of this material by Kate Scott strongly suggests that human hunters used the ravine at the end of a headland to drive groups of mammoth and woolly rhinoceros to their deaths. The restriction of the 'bone piles' to only two horizons, however, must indicate that animals were not killed in this fashion regularly, but probably when these large beasts happened to wander on to the headland. Alternatively it is also possible that they blundered over the precipice without any human assistance.

Evidence for human presence during much of the succeeding Ipswichian Interglacial, about 110,000 to 120,000 years ago, is more abundant than for the Wolstonian, but there are still few artefacts.

A few flint flakes from Selsey, West Sussex, are almost the only evidence of human presence from the early part of the Ipswichian (probably pollen zone I). At Barrington, near Cambridge, pits worked around 1900 yielded spectacular remains of hippopotamus, straight-tusked elephant, bison, aurochs, giant deer, red deer, fallow deer, lion, spotted hyena and other mammals. This highly characteristic assemblage of fossil mammals, known as the 'hippopotamus fauna', is recorded from numerous localities throughout England and Wales, for example Trafalgar Square

(London), Victoria Cave (North Yorkshire) and Joint Mitnor Cave (Devon). Where associated fossil pollen has been found these faunas have been dated to late pollen zone II and early zone III. Forest cover included oak, elm, alder, maple, birch, pine, lime and hazel. Very high levels of herb pollen at some sites, such as Barrington, are thought to have resulted from local destruction of the forest by the large herbivorous mammals. Horse, so common at other times, is curiously absent from these 'hippopotamus faunas'. Humans were almost certainly also absent; there is no record of artefacts from this period, although conditions appear to have been very suitable. Geological evidence, however, suggests that Britain had become an island for the first time in the ice age, so it may have been the narrow forerunner of the English Channel which was responsible for excluding people.

Artefacts and mammalian remains occur in several sites assigned to the second half of the Ipswichian Interglacial (pollen zones III and IV). Many workers believe, however, that the faunas and archaeological material date from a hitherto unrecognised interglacial between the Hoxnian and Ipswichian (chapter 2). Man's role in accumulating animal remains is not clear at any of these sites. At Stutton and Brundon, Suffolk, flint flakes made by Levalloisian technique occur in association with mammoth, straight-tusked elephant, horse, extinct rhinoceros *Dicerorhinus hemitoechus,* pond tortoise and other species. At Crayford, Greater London, sands and silts of an earlier river Thames yielded Levalloisian blades and Acheulian handaxes in association with a fauna including mammoth, horse, red deer, woolly rhinoceros, the 'interglacial' rhinoceros *D. hemitoechus,* lion, bear and wolf. The age of these deposits is enigmatic, but they may date from the very end of the Ipswichian Interglacial (zone IV?). A rather similar fauna is known from Stoke Tunnel/ Maidenhall, Ipswich, Suffolk, in association with sparse artefacts. Here the presence of pond tortoise emphasises that fauna and artefacts date from a period of temperate climate.

Artefacts of Mousterian type, associated elsewhere with Neanderthal man, occur sparsely in southern Britain in deposits apparently dating from the first two thirds or so of the Devensian/Last Cold Stage. In the lower part of the cave earth at Kent's Cavern, Torquay, Devon, they appear to occur in broad association with mammoth, woolly rhinoceros, horse, reindeer and other animals common in faunas of Devensian age, accumulated wholly or partly by spotted hyenas and broken and chewed by them. Such damage obliterated any possible evidence of

21. The rugged limestone country around Victoria Cave, North Yorkshire. Although at latitude 54 50N and 440 metres (1444 feet) above sea level, this cave has yielded remains of hippopotamus, spotted hyena and other animals dating from the Ipswichian/Last Interglacial. (Photograph: author.)

human hunting or butchering. There is no precise information as to the age of any Mousterian artefacts from Britain. It is evident, however, that the climatic conditions of the early and middle Devensian were generally milder than later in the stage. Even so, the scarcity of artefacts suggests that humans may have been present only during the warmest (interstadial) phases.

Finds of skeletal remains from a number of localities in Europe and elsewhere show that Neanderthals were shorter and considerably more robust and muscular than modern man. The large skull was long and low with massive brow ridges, a sloping forehead, a projecting face with swept-back cheek bones, and a brain as large or larger than modern man's. The large jaw usually lacked a chin, but in some specimens this feature (especially characteristic of modern man) is weakly developed.

7
Advanced hunters

The Devensian or Last Cold Stage was a period of cold and rather dry climatic conditions, accompanied by herbaceous vegetation comprising mainly grasses and sedges. These prevailing conditions were interrupted at intervals by warmer interstadials and much colder phases (see chapter 3). The interstadials were usually marked by the growth of birch or conifer woodland. Two such episodes (the Wretton and Chelford Interstadials) have been recognised in the early part of the Devensian, while an interstadial in the later part of the Devensian (Late Glacial Interstadial), lasting from approximately 14,500 to 10,600 years ago, was accompanied by the growth of birch woodland and, in south-east England, possibly also pine. During the coldest phases, 'polar desert' (arctic conditions with sparse vegetation) appears to have been present. Within such a phase, which lasted from about 25,000 to 14,500 years ago, occurred the last major glaciation of the British Isles, about 18,000 to 15,000 years ago (figure 1). Following the Late Glacial Interstadial there was a brief but sharp return to arctic conditions (Loch Lomond Stadial or 'Younger Dryas' period) between about 10,600 and 10,000 years ago, resulting in local glaciation of north-west Scotland and other highland areas.

People of modern type ('anatomically modern' humans) arrived in western Europe about 35,000 years ago. As discussed more fully in chapter 8, they appear to have been more advanced than their Neanderthal predecessors, whom they displaced. This is shown by the more varied and sophisticated range of tools and weapons manufactured, referred to collectively as 'Upper Palaeolithic' (late old stone age) industries. In continental Europe, where artefacts are more abundant and recorded from essentially continuous sequences, a whole series of names is employed to describe the different Upper Palaeolithic assemblages (for example Aurignacian, Solutrean, Magdalenian of the classic French sequence). Because the artefacts from Britain are sparse and often lack exact parallels elsewhere, it is convenient to use the simple division proposed by J. B. Campbell: 'Early Upper Palaeolithic' (EUP) and 'Later Upper Palaeolithic' (LUP). The age range is now taken as at about 36,000 to 35,000 years ago for EUP and about 12,500 to 10,000 years ago for LUP; the gap in time between them corresponds to the 'polar desert' and glaciation phase, when Britain was presumably uninhabitable.

Most Upper Palaeolithic finds in Britain are from caves, and dating usually relies solely on radiocarbon measurements of fossil mammal bones. Moreover cave sediments do not contain well preserved fossil pollen, so that past environments usually have to be reconstructed from the vertebrate remains present at these and other sites, and from fossil plant remains preserved elsewhere.

Picken's Hole, a tiny cave in the Mendip Hills, Somerset, yielded a few nondescript flint flakes, two human teeth of modern type and a rich assemblage of mammal remains datable to about 27,540 and 34,265 years ago. The fauna includes spotted hyena, lion, bear, arctic fox, mammoth (baby individuals), woolly rhinoceros, horse, reindeer, extinct bison and ground squirrel, which indicates open vegetation rich in grasses with few trees; consistent with a climate colder and drier than today. The site was too small for long-term human occupation and may record a single visit by one or more persons seeking temporary shelter. As at many such cave sites in Britain, most of the remains, including those of humans, appear to have been brought in by spotted hyenas (see chapter 3). Very similar faunas are known, for example from Kent's Cavern, Devon, broadly in association with radiocarbon dates ranging from about 38,270 to 27,730 years ago, and from several of the caves in the Creswell Crags on the Derbyshire/Nottinghamshire border, Ffynnon Bueno Cave, Clwyd, North Wales, and King Arthur's Cave (Hereford and Worcester) — in each case in association with EUP artefacts. Such industries produced flint points almost certainly used to tip spears; bone and antler spear points of the same period are common on the European mainland. So far, British sites have yielded little direct evidence for the way of life of EUP peoples. This is partly because none of the best sites has been excavated to a high standard, but also because it is not possible to distinguish animal bones brought in or damaged by humans in the presence of overwhelming damage inflicted by the powerful jaws and teeth of hyenas. Mammal remains apparently accumulated by man from some large rock shelters ('abris') of the Dordogne, France, in the earlier part of the Upper Palaeolithic period indicate that a wide range of species was hunted. These included horse, reindeer and red deer — all of which would also have been available in southern Britain — as well as wild boar, roe and chamois, which did not then range so far north.

There is fortunately rather more information available on the

22. (Left) Cheddar Gorge, Somerset, cuts spectacularly into limestone of Carboniferous age. The gorge contains many caves, the largest being Gough's Cave, which have yielded the bones and teeth of animals together with the tools and weapons made by Upper Palaeolithic (late old stone age) people. (Photograph: Cadbury Lamb.)

23. (Below) King Arthur's Cave (Hereford and Worcester). An important Upper Palaeolithic (late old stone age) site which contained a sequence of layers, each with artefacts and animal remains. (Photograph: author.)

24. Ffynnon Bueno Cave, Clwyd, North Wales. This and the neighbouring Cae Gwyn Cave (below the fence on the left of the photograph) when excavated in the nineteenth century yielded Upper Palaeolithic (late old stone age) artefacts in association with remains of spotted hyena, reindeer, horse, mammoth and other animals. The cave entrances were sealed by boulder clay, a glacial deposit, proving that the finds dated from before the last glaciation of the area between 17,000 and 18,000 years ago. (Photograph: author.)

LUP peoples who reoccupied Britain after the retreat of the ice sheet, in the period known as the Late Glacial. LUP humans were, however, living in a slightly, but perhaps significantly, different environment in that there had been some major changes in the animal world. In western Europe mammoth and woolly rhinoceros appear to have become extinct by about 12,000 years ago, although mammoth may have survived rather later in Siberia (see chapter 9). Both mammoth and woolly rhinoceros, however, are recorded from a few sites which postdate the main glaciation, notably the remains of an adult mamoth and three juvenile from Condover, Shropshire, dated to about 12,800 years ago (see figure 35). At Gönnersdorf, West Germany, engravings of these and other mammals, as well as fossil remains, were discovered at an Upper Palaeolithic site dated to about 12,500 years ago.

In 1970 the entire skeleton of an elk (moose) was found within the sediments of a former lake on the outskirts of Blackpool, Lancashire. Studies of fossil pollen showed that the sediments had been deposited during the later part of the Devensian/Last Cold Stage and that the elk dated from the milder phase (Late Glacial Interstadial) that occurred within this period. A date of about 12,400 years ago was obtained by the radiocarbon method.

Of particular interest was the discovery during excavations by J. S. Hallam and B. J. N. Edwards of two barbed spear points made of antler in association with the skeleton. One was found resting in a groove in the end of the hind cannon bone (metatarsal). This groove had resulted from infection of a wound caused by the point which was embedded in the animal's foot one or two weeks before its death. Examination by the author of the rest of the skeleton revealed that many bones bore signs of damage, probably by stone-tipped weapons, which had occurred immediately before death. Notwithstanding all this evidence of human activity, however, the skeleton was found intact — it had not been butchered and eaten. A reasonable interpretation of the find is that a hunting party away from their base attacked an elk, probably deliberately disabling it by embedding a harpoon point in its hind leg. They then either followed or drove the animal some distance, intending to kill it at a more convenient spot. However, although seriously wounded in a second attack, it escaped the hunters only to drown in a lake. A further detail is that the animal's antlers were about to be shed, indicating that all this took place in the winter. Modern elk sometimes die when they break through ice on lakes and rivers and are unable to get out again. This may well have been the fate of the Blackpool animal, especially as it would have been weakened by injuries.

At Gough's Cave, in the Cheddar Gorge, Somerset, excavated in the 1920s and 1930s by R. F. Parry, LUP artefacts were found in association with animal bones. The latter, studied by A. P. Currant, mostly comprise horse together with red deer, reindeer, saiga antelope, arctic hare, pika (a small short-eared relative of hares and rabbits, now confined to central and eastern Asia), wolf, arctic fox, bear, beaver, water vole and Norway lemming. A series of radiocarbon dates indicates that this assemblage probably dates to between 12,800 and 11,900 years ago, that is from within the relatively mild Late Glacial Interstadial period.

Many of the bones have cut marks made by stone tools, and the assemblage appears to consist of the remains of animals killed by humans. In addition to flint artefacts the site has also yielded objects made from bone and antler, including a perforated piece of reindeer antler (a so-called *baton de commandement*), probably used for straightening the shafts of spears and perhaps arrows, fox-tooth beads, piercers made from the bones of hare and pieces of large-mammal bone from which needles had been cut. Kate Scott has suggested that as horses are represented largely by foot bones, and that cut marks (made by stone tools)

25. Reconstructed scene in southern England about 30,000 years ago during the Devensian/Last Cold Stage. A treeless steppe-tundra vegetation supported herds of mammoth, horse, bison and reindeer. Early Upper Palaeolithic hunters are shown defending their 'kill' from spotted hyenas. (The use of a spear-thrower at this early period is likely, but conjectural.) (Line drawing by N. Arber.)

are found on the undersides of their toe bones (phalanges), long tendons were obtained by carefully cutting up the feet from butchered horses. Such strong and flexible thread could have been put to many uses, for example sewing hides and binding a spear head to its shaft. Butchery of horse and red deer at the site has been described by Parkin, Rowley-Conwy and Serjeantson. From analyses of cut marks they concluded that feet of both horse and red deer were processed for tendons, and that carcasses were dismembered and stripped of meat in the cave. This site demonstrates the importance of animals to early man as a source of raw materials.

Another very informative Upper Palaeolithic site is the very small limestone Ossom's Cave, in the Manifold valley, Staffordshire, excavated by D. Bramwell in the 1950s. The animal remains have been studied recently by Kate Scott, who has pieced together a fascinating picture from life in the closing phases of the ice age. The principal bed, 'Layer C', consisted of angular limestone fragments, split off by frost action from the cave roof, and contained abundant remains of reindeer, together with fragments of horse and perhaps bison in association with flint

26. The leg bones of a fossil elk skeleton under excavation near Blackpool, Lancashire, in 1970. The bones are seen lying in a series of clays and peaty sediments deposited in a former lake about 12,500 years ago. (Photograph: J. S. Hallam.)

27. (Below) A more advanced stage of the excavation shown in figure 26. A barbed spear point can be seen as found resting (in a groove that it had worn) at the lower end of the left hind cannon bone, snapped in two by the pressure of the overlying sediments. (Photograph: J. S. Hallam.)

28. Caves in the Manifold valley in the Staffordshire Peak District. The prominent cave to the left of the picture, Thor's Cave, appears to have contained no palaeolithic (old stone age) material, and all of the deposits were removed in the nineteenth century. Several other smaller caves in the same crag have yielded remains of ice age mammals and artefacts. (Photograph: M. J. Bishop.)

LUP artefacts. A concentration of ash at the cave mouth showed where fires had been lit. Small mammals present included arctic lemming and Norway lemming, which together with reindeer indicate a severe climate, but also bank vole which points to milder conditions. Radiocarbon dates indicate that this assemblage dates from about 10,600 to 10,780 years ago, that is from the latest part of the mild Late Glacial Interstadial, just before the onset of the renewed cold period (Loch Lomond Stadial) at the end of the Devensian.

The fossil remains represent at least six reindeer, including both young and mature individuals, the bodies of which must have been dragged up into the cave. Most bones show signs of butchery, which compare closely with the techniques for cutting up reindeer carcasses used by Eskimos. In particular the ends of some limb bones show characteristic damage, and the large numbers of bone splinters are identical to those produced by Eskimos in cracking reindeer bones for their fat-rich marrow. Because reindeer are born around mid May, and young animals can be accurately aged from the stage of eruption of their permanent teeth, it is possible to determine the season of death rather precisely if, as at Ossom's Cave, jaws of juveniles are present. From this evidence it was apparent that these reindeer had been killed in the spring. A vivid picture emerges of hunters using the tiny cave as a temporary shelter and lookout station, ambushing the reindeer herds as they migrated from lowland

29. (Above) Ossom's Crag in the Staffordshire Peak District, containing Ossom's Cave, which was completely excavated in the 1950s. When occupied by reindeer hunters, towards the end of the ice age, it provided both temporary shelter and a commanding view of the Manifold valley. (Photograph: M. J. Bishop.)

30. (Left) The entrance to Ossom's Cave. (Photograph: author.)

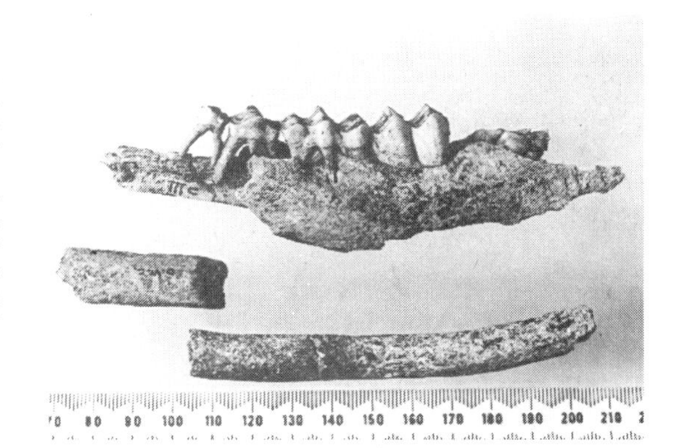

31. Remains of young reindeer from Ossom's Cave, Staffordshire. The partial lower jaw (mandible) is from one of several individuals killed when they were about one year old, that is, in the spring. Corroborative evidence for the season when the reindeer were hunted is given by the antler fragments, also from animals which died in their first year. (Photograph: K. Scott.)

wintering grounds up the Manifold valley to summer pastures in the Peak.

At Sproughton near Ipswich, Suffolk, barbed spear points made of antler were found in the same deposits as remains of horse and reindeer, although not in direct association. Radiocarbon dates of about 10,910 and 10,700 years ago on the barbed points again indicate human presence at the end of the Late Glacial Interstadial.

Claims have been made for the presence of people at the end of the Devensian well beyond the area of southern Britain, based on collected and worked reindeer antlers from a cave at Inchnadamph in the far north-west of Scotland. The remains show every indication of having been accumulated by wolves, however, not humans. Moreover there are no convincing records of palaeolithic artefacts further north than Kirkhead Cave, Cumbria.

Although strictly beyond the scope of this book, because hunting and gathering continued in Britain well beyond the nominal end of the ice age, 10,000 years ago, into the first half of the Flandrian or Postglacial period in which we now live, it is appropriate to comment briefly on the way of life of middle stone age (mesolithic) people. In the 1950s excavations directed by J. G. D. Clark at Star Carr, near Pickering, North Yorkshire, uncovered the remains of an encampment beside a former lake, dated by radiocarbon to about 9500 years ago. On the evidence of fossil pollen, birch and pine woods grew over much of the region with local areas of grass and herbs and reedswamp fringing the shore of the lake. Numerous artefacts, including barbed spear points made of antler (probably for hunting large mammals) and specialised arrowheads (presumably for killing birds and small

32. A series of caves near Inchnadamph, north-west Highlands of Scotland. Despite claims to the contrary, these caves have not produced good evidence for the presence of Upper Palaeolithic (late old stone age) people in Scotland, although reindeer remains were found in abundance. (Photograph: author.)

mammals), were found in association with many animal bones and teeth. Species taken included red deer, elk (moose), aurochs, roe deer, wild boar and various waterfowl. Dogs were probably used in hunting. There is ample evidence from sites of similar age ('Maglemosian' culture) on the North European Plain for fishing using hooks and nets, but none from Star Carr. Later mesolithic groups in Scotland are known to have extensively exploited marine resources, especially shellfish, as well as hunting mammals and birds.

People living in Britain at this time enjoyed a relatively highly developed technology and the wide range of resources available in a period of temperate climate. This potent combination enabled humans with mesolithic cultures to colonise northern Britain and Ireland — regions previously uninhabited. After about 6000 years ago, however, the first new stone age (neolithic) farmers, bringing with them their domestic plants and animals, arrived from the European mainland. This new way of life, with its assured food supply and the resulting much higher human population densities, rapidly replaced the uncertainties of hunting and gathering and in turn led to the development of civilisation and our progressive alienation from the natural world.

8
Making a living

Humans apparently originated in Africa, where the earliest known fossils that can be assigned to genus *Homo* date back to well over a million years ago. People did not reach Europe until later, and Britain, on the northern fringes of the palaeolithic world, was not colonised until 350,000 to 400,000 years ago, at the end of the Cromerian Interglacial. As can be seen from figure 33, the record of man in Britain for most of the ice age is intermittent. There is a strong possibility that until the appearance of peoples with Upper Palaeolithic cultures, about 35,000 years ago, humans were present in Britain only when climatic conditions were least harsh and there was the greatest range of animal and plant foods available.

It is impossible to assess past population numbers or population densities accurately, but it is apparent from the relatively sparse numbers of artefacts recorded from sites in Britain, compared with their abundance further south in Europe, that even when people were in Britain their numbers were low. By analogy with today's hunter-gatherers, a reasonable estimate of the maximum number of people occupying Britain during the ice age might be a few hundred, at the most a few thousand. Palaeolithic artefacts, of any period, are not recorded further north than Cumbria and there is no convincing record of man from Scotland or Ireland until after 9000 years ago, over a thousand years after the end of the ice age. The overall impression is that throughout the ice age conditions in Britain were marginal for human existence; thus the area is of considerable interest in understanding how early man adapted to his environment.

It is remarkable how the possession of even simple technologies enabled man to expand far beyond his original geographical range. One major factor was the use of fire, as indicated by burnt flints, as at Hoxne and Pontnewydd, and ash remaining from hearths at a number of later cave sites. Further protection from the cold was provided by clothing. Eyed bone needles, presumably used to sew skins, are known from a number of Upper Palaeolithic sites, especially in France. More spectacularly, Upper Palaeolithic human skeletons found at Sungir near Moscow were covered with beads clearly once sewn on to a kind of anorak and trousers.

Palaeolithic peoples made use of caves for both temporary and more permanent shelter, but caves in Britain are few and restricted to particular areas, usually with limestone bedrock. The ability to construct huts or tents must have been essential for survival. Although so far there are no such finds from Britain, a number of Upper Palaeolithic sites in continental Europe, for example in Germany and northern France, suggest that dwellings, some perhaps portable, consisted of animal hides on a framework of wooden poles. Huts constructed entirely of mammoth bones are known from the Ukraine, and other sites evidence dwellings which appear to have had a covering of hides weighted at the edges with tusks, bones and rocks. In interglacials and interstadials when trees were present in abundance, huts must have been built from branches, perhaps with roofs of brushwood or even thatch.

Perhaps the most important factor to consider is the quest for food. Direct evidence for human diet is sparse and difficult to interpret. However, the deeply entrenched idea that from earliest times humans were active and successful hunters of even the largest and most formidable animals has been questioned. The mere association of artefacts and animal bones in a deposit is no proof that the animals represented were killed by man (chapter 3). Peoples with Lower Palaeolithic (early old stone age) cultures, living during the Hoxnian Interglacial, left handaxes and other artefacts at such sites as Hoxne and Swanscombe, sometimes associated with quantities of mammal remains which could be human food refuse. But do these remains represent animals actively hunted by man, occasional opportunist killings of sick or injured individuals, or remnants scavenged from the kills of carnivores such as lions, wolves or hyenas? Could humans have been efficient hunters at this early period? They almost certainly lacked weapons which could be used at a distance, such as the throwing spear and bow and arrow.

This situation was little changed approximately 100,000 years later, when rather more sophisticated artefacts classified as Middle Palaeolithic (middle old stone age) were being manufactured by early humans, including Neanderthals. There is no evidence that the stone points of 'Mousterian' or 'Levalloisian' type were hafted to tip throwing spears, although they may have been so used. In any case the technologies of that period were evidently much inferior to those of the Upper Palaeolithic (late old stone age) which followed. During part of the Wolstonian Cold Stage it is probable that hunters at La Cotte de St Brelade, Jersey, deliberately stampeded mammoths and woolly rhinoceros

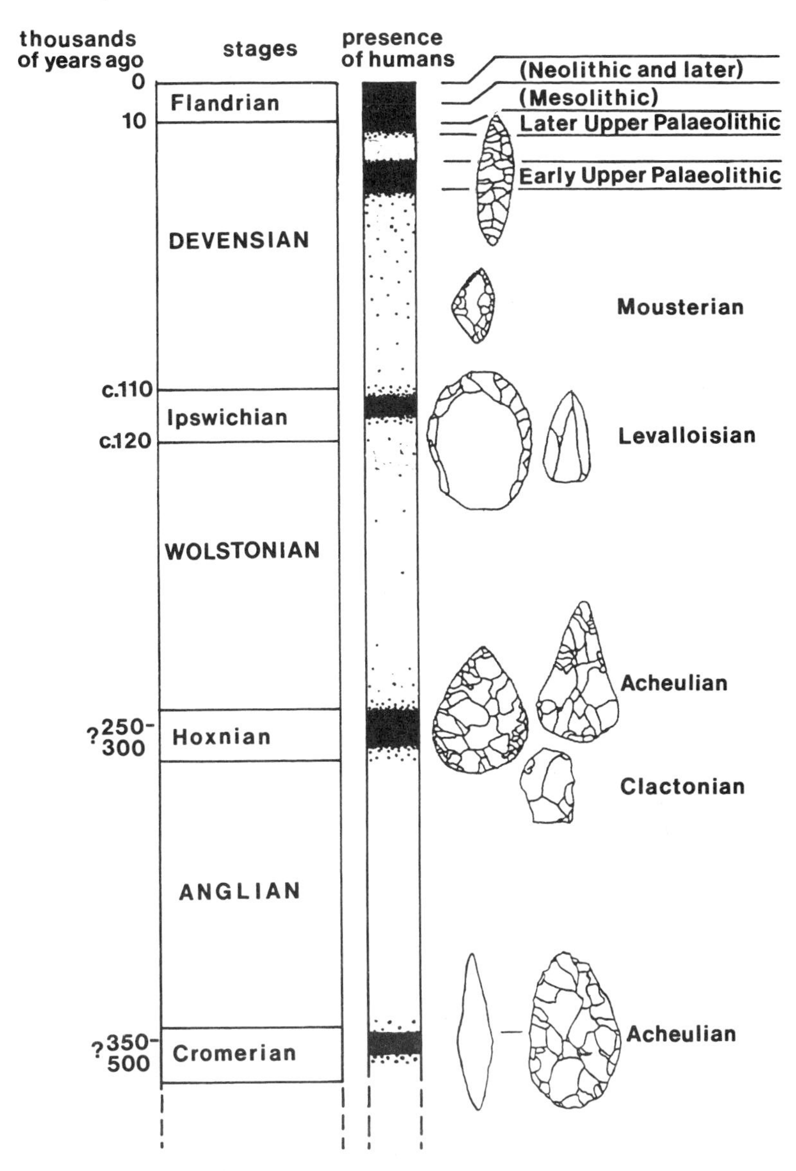

33. Record of human presence in Britain during the ice age. The intensity of shading reflects the degree of certainty of human presence at a particular time. Representative artefacts (not to scale) are shown: Acheulian handaxes; Clactonian flake; Levalloisian core and point; Mousterian point; Early Upper Palaeolithic leaf-shaped spear point.

into a ravine (chapter 6).

The fossil evidence from both archaeological and non-archaeological sites of Hoxnian age shows that people living in southern Britain at this time would have had access to a wide variety of potential foods. These would have included various fruits and berries, hazel nuts, fungi, certain plant roots, leaves such as nettles and fat-hen, birds' eggs, frogs, snails, freshwater mussels and other invertebrates and fishes in addition to meat from mammals. Even though these items were available there are no means of determining which were actually eaten. Fishes (marine as well as freshwater) are sparsely represented at Lower and Middle Palaeolithic sites, and it seems probable that fishing was not practised on any scale. Convincing evidence for the use of plant foods awaits the possible discovery of an archaeological site with plant material preserved in sediments, indicating stored provisions or food refuse. Groups living near the sea must have had access to seaweeds and marine shellfish, and the occasional stranded whale, dolphin or porpoise, but again no evidence has so far been found.

Many of these foods would have been available only seasonally and, again by analogy with modern hunter-gatherers, some could have been preserved; meat by sun drying or smoking; fruits and berries by being dried or stored in a container and allowed to ferment to produce an alcoholic mash.

Discoveries south of the Sahara suggest that 'anatomically modern' humans (with skeletal features much as in present-day people) may have originated in Africa. They reached western Europe about 35,000 years ago, replacing the Neanderthals occupying that region, some five thousand years later than the earliest records from the Middle East. Where fossil remains of anatomically modern people (also known as 'Cro-Magnon' after the classic limestone rockshelter site in the Dordogne, France) occur in association with artefacts, these are invariably of Upper Palaeolithic (late old stone age) type. It is clear from the archaeological record that something momentous had occurred in human prehistory. The striking feature of the Lower and Middle Palaeolithic is the incredibly slow pace of technological change, the same kinds of artefacts being found in deposits separated by tens of thousands of years and hundreds or even thousands of kilometres.

In contrast Upper Palaeolithic artefacts show much more variation, even differing markedly from neighbouring regions. Moreover they changed rapidly with time and archaeologists use

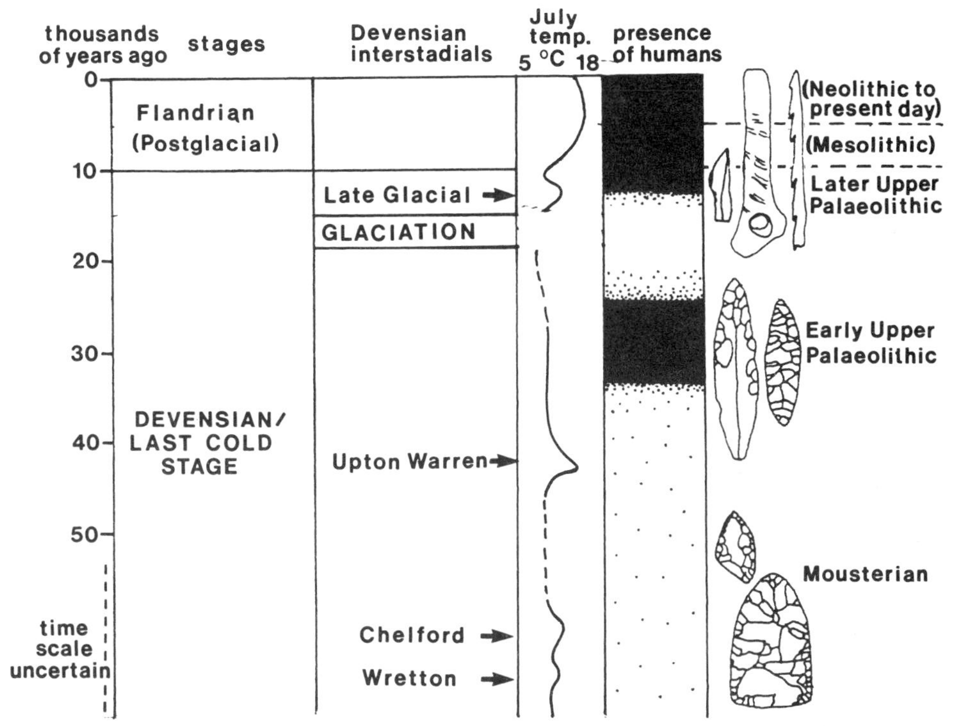

34. Record of human presence in Britain during the Devensian Last Cold Stage, in more detail than in figure 33. The representative artefacts are not to scale: Mousterian hand axe and point; Early Upper Palaeolithic leaf-shaped spear point; Later Upper Palaeolithic flint spear point, 'baton' (shaft straightener?); and barbed spear point.

a multiplicity of names to describe the many different cultures. Studies by Professor R. Klein of animal bones from sites in southern Africa indicate that late stone age peoples (who arrived there about 20,000 years ago) were more efficient hunters than middle stone age cultures, exploiting a wider variety of available animals, including such dangerous prey as wild pigs. The same general pattern was probably also true for Britain, and the overall impression is that Upper Palaeolithic humans were of greater intelligence, organisation and innovative ability than any that had gone before. Their 'tool kits' were also relatively very sophisticated, including objects made of bone and antler as well as of stone. Throwing spears were used extensively as is shown by abundant finds of what are clearly spear points made of flint, and plain and barbed forms made of bone and antler. Finds from the

Upper Palaeolithic of France show that spear-throwers, providing an enormous increase in both force applied and range, were in use at least by about 17,000 years ago (Magdalenian industry). Artefacts from the Upper Palaeolithic of Spain (Solutrean industry, about 20,000 years ago) looking exactly like arrowheads suggest that the bow and arrow may have been invented as well.

Upper Palaeolithic people were also the first to make art objects, including animal and human figures carved out of bone and ivory as well as the celebrated cave paintings from France and Spain. Unfortunately no cave paintings and few carved objects are known from Britain, possibly because art for ceremonial or ritual purposes was not appropriate for the transitory life in Britain, in contrast to the more permanent occupations further south.

Significantly the Upper Palaeolithic in Britain, as elsewhere, dates entirely from within a period of predominant cold — the later part of the Devensian/Last Cold Stage. Presumably only the possessers of this relatively advanced technology, and the skills and organisation to go with it, were able to live as far north as Britain during a cold phase of the ice age. It is also apparent from fossil evidence that few plants or invertebrates would have been available during cold phases. So Upper Palaeolithic peoples living in Devensian Britain must have subsisted on an essentially all-meat diet (as did Eskimos until recently), and were therefore efficient hunters. There is good evidence that horses, reindeer and elk (moose) were successfully hunted in Britain (chapter 7). No doubt other mammals and perhaps birds were also taken. There is so far no evidence for hunting of mammoth, woolly rhinoceros or other very large mammals. Possibly human groups migrated into Britain only for the spring and summer to take advantage of seasonally available supplies of game. It is widely thought that they followed the constantly moving reindeer herds on their seasonal migrations of hundreds of kilometres. It should be stressed, however, that the elk skeleton with associated artefacts from Lancashire (chapter 7) shows unequivocally that humans could survive through the winter even in northern England. Moreover, reindeer herds commonly cover as much as 65 km (40 miles) a day and it is hardly credible that human groups, encumbered as they would have been with babies and the old or sick, would have been able to keep up.

Even Upper Palaeolithic peoples were not able to live in Britain during the maximum of the last glaciation, about 15,000 to 18,000 years ago. Moreover, they could not colonise Scotland

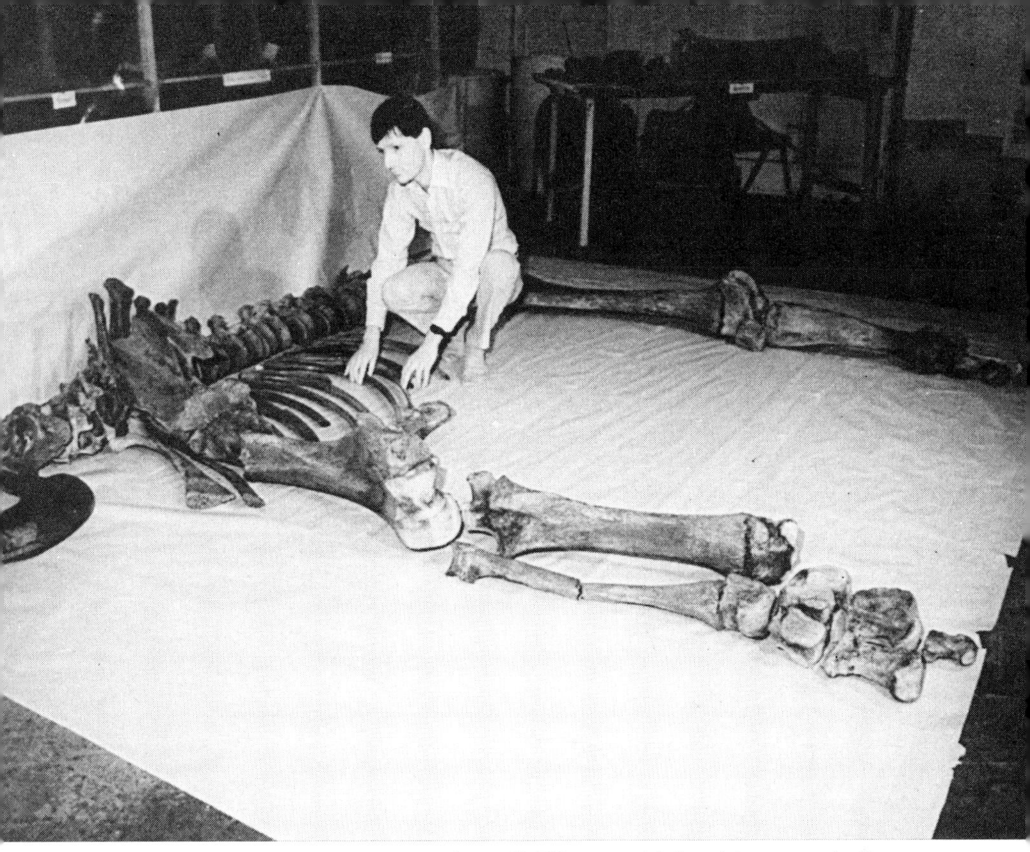

35. Skeleton of an adult mammoth about 12,800 years old found in a gravel pit at Condover, Shropshire. The limb bones of the leftside only are shown here; the skull was not recovered. (Shropshire Museums Service, photograph by Colin Shuttleworth).

or Ireland (for which boats would probably have been needed) and it was only after about 9000 years ago, well after the end of the ice age, that mesolithic (middle stone age) hunter-gatherers spread over the entire British Isles.

9
Overkill

The disappearance of many species of large mammals at the end of the ice age, throughout most parts of the world, is one of the most intriguing mysteries in the history of life on earth. By at the latest 10,000 years ago, for example, North America had lost about 73 per cent of its large mammals — with adult body weights over about 40 kg (88 pounds), including mastodons, giant ground sloths, giant armadillo-like animals and a sabre-tooth cat. In Australia, with a much smaller fauna to begin with, extinctions were even more spectacular, affecting several species of giant kangaroo, a wombat-like animal as big as a rhinoceros (*Diprotodon*) and the 'marsupial lion' (*Thylacoleo*).

The same phenomenon occurred also in Europe, although here it was less marked, affecting about 30 per cent of large mammals. As described in chapter 7, the fauna of Britain and Europe about 30,000 years ago, towards the end of the Devensian/Last Cold Stage, included mammoth, woolly rhinoceros, giant deer and cave bear, all of which were entirely extinct by 10,500 years ago. Moreover, straight-tusked elephant and the rhinoceros *Dicerorhinus hemitoechus*, common in Britain during the Ipswichian/Last Interglacial, about 120,000 years ago, also disappeared. *D. hemitoechus* survived in southern Europe until at least 20,000 years ago, but the straight-tusked elephant may have died out everywhere much earlier. In addition, several species, although alive today, suffered drastic reductions in geographical range. Hippopotamus, so characteristic of Britain during the Ipswichian/Last Interglacial, failed to return when the climate warmed up again in the Flandrian/Postglacial and is today restricted to Africa. The spotted hyena, which evidently thrived in Britain in both the Ipswichian Interglacial and the Devensian/Last Cold Stage, is also now found only in Africa, while the lion, with a long Pleistocene history in Britain and Europe, was historically restricted in Europe to a small area of the Balkans.

Extinctions occurred earlier in the ice age, but they affected small mammals as much as large ones and were spread out over a much longer period of time. Unlike earlier phases of 'mass extinction' in the geological record, for example the major event at the end of the Cretaceous Period about 66 million years ago affecting dinosaurs on land, flying reptiles in the air and ammonites and various marine reptiles in the sea, the extinctions at the end of the ice age were with few exceptions confined to large terrestrial mammals.

What caused the demise of the mammoth, woolly rhinoceros, mastodon, giant kangaroos and the other large mammals across the globe? Catastrophic theories, such as asteroid impact, proposed by some to explain the extinction of the dinosaurs, are not appropriate for the ice age because extinctions were not simultaneous everywhere. In Europe, for example, mammoths appear to have gone by about 12,000 years ago, whereas the giant deer survived as late as 10,500 years ago. Extinctions were clearly staggered over a long period. In North America, however, mastodons, mammoths and many other extinct mammals all lasted much longer and their disappearance was a relatively sudden event.

There are two serious contending hypotheses for explaining the great ice age extinctions, a topic which attracts a great deal of interest. Some believe extinctions resulted when animals were unable to adapt to climatic changes, while others, notably Professor P. S. Martin of the University of Arizona, think they were exterminated by human hunters — 'prehistoric overkill'. In either case it is the larger, slowest-breeding species which are generally supposed to have been the most vulnerable.

Workers who favour climatic causes point to the rapid and marked changes in climate and vegetation that occurred over most of the earth towards the end of the ice age. From about 13,000 years ago the open vegetation rich in grasses and herbs which covered vast areas of Eurasia and North America began to give way to forests, with the onset of warmer and wetter conditions. The extinction of mammoth and woolly rhinoceros is supposed to have resulted from the loss of their habitat. The arguments against this are twofold. Firstly, many animals widespread in now temperate regions during the Last Cold Stage, such as reindeer, lemming and arctic fox, retreated to the arctic at the end of the ice age, while others such as ground squirrels and saiga antelope moved eastwards to the grassy steppes of central Asia. Why did not the mammoth, woolly rhinoceros and other extinct species respond similarly by migrating to other suitable areas? The second difficulty is that profound changes in climate were a recurring feature of the ice age, and previous similar transitions from cold to warm (interglacial) periods were not accompanied by mass extinction.

The alternative 'overkill' hypothesis, however, also fails to account for all the observed facts. If this were the cause, one would expect extinctions to have occurred following the first arrival of humans in an area or the introduction of a new (Upper Palaeolithic, or late old stone age) hunting technology. This appears to obtain in North America, where extinctions seem to have coincided closely in time with the arrival of hunters of the Clovis culture about 11,000 years ago, but in Australia human arrival very probably occurred tens of thousands of years before extinctions. In Europe, too, most extinctions happened long after the appearance of Upper Palaeolithic cultures. Moreover, it is difficult to imagine how a few hunters, with primitive technology, could have exterminated so many species, including many formidable large mammals, over their entire geographical ranges.

Much work needs still to be done on this, but the answer may well be a combination of the two factors: man and climate. Towards the end of the ice age climatic changes would undoubtedly have restricted animals with formerly extensive ranges. Although this had happened several times previously at the onset of interglacials, this contraction of geographical area was not sufficient to cause 'mass extinction'. The new factor, however, was perhaps the presence of humans equipped with relatively advanced hunting weapons and skills. If so, the long history of interaction of humans and animals throughout the ice age culminated in mass exterminations, a salutary demonstration of the unwitting power of man to modify and even destroy his environment.

10
Further reading

Lowe, J. J., and Walker, M. J. C. *Reconstructing Quaternary Environments*. Longman, 1984

Mellars, P. A. 'The Palaeolithic and Mesolithic', in Renfrew, C. (editor), *British Prehistory; a New Outline*. Duckworth, 1976.

Stuart, A. J. *Pleistocene Vertebrates in the British Isles*. Longman, 1982.

Sutcliffe, A. J. *On the Track of Ice Age Mammals*. British Museum (Natural History), 1985.

Timms, P. *Flint Implements of the Old Stone Age*. Shire Publications, second edition, 1980.

West, R. G. *Pleistocene Geology and Biology; with Special Reference to the British Isles*. Longman, second edition, 1977.

Wymer, J. J. *The Palaeolithic Age*. Croom Helm, 1982.

11
Places to visit

Many museums in Britain have some ice age material on display. The following deserve particular mention.

British Museum (Natural History), Cromwell Road, South Kensington, London SW7 5BD. Telephone: 01-938 9123. Extensive displays, including finds from Swanscombe and other Thames deposits, Devonshire caves, etc; also important non-British material relevant to ice age extinctions.

Buxton Museum and Art Gallery, Terrace Road, Buxton, Derbyshire SK17 6DJ. Telephone: Buxton (0298) 4658. Peak District cave material.

Cambridge University Museum of Archaeology and Anthropology, Downing Street, Cambridge CB2 3DZ. Telephone: Cambridge (0223) 337733. Finds from the mesolithic site of Star Carr, North Yorkshire.

Castle Museum, Norwich, Norfolk NR1 3JU. Telephone: Norwich (0603) 611277, extension 279. Fossil mammals from the Cromer Forest Bed, Norfolk and late Pleistocene sites in Norfolk.

Cheddar Caves Exhibition and Museum, Cheddar Caves, Cheddar, Somerset BS27 3QF. Telephone: Cheddar (0934) 742343. Finds from Gough's Cave; guided tour of Gough's Cave.

City of Bristol Museum and Art Gallery, Queen's Road, Bristol BS8 1RL. Telephone: Bristol (0272) 299771. Material from Mendip Caves and gravels of the river Avon; mounted giant deer skeleton.

Creswell Crags Visitors' Centre, Crags Road, Creswell, Worksop, Nottinghamshire. Telephone: Worksop (0909) 720738. Creswell Crags material; guided tour of caves.

Cromer Museum, East Cottages, Tucker Street, Cromer, Norfolk. Telephone: Cromer (0263) 513543. Small display of finds from Cromer Forest Bed.

Harris Museum and Art Gallery, Market Square, Preston, Lancashire PR1 2PP. Telephone: Preston (0772) 58248. Mounted skeleton of elk from Blackpool.

Ipswich Museum, High Street, Ipswich, Suffolk IP1 3QH. Telephone: Ipswich (0473) 213761/2. Material from Brundon and other important Suffolk localities.

Leicestershire Museum and Art Gallery, 96 New Walk, Leicester LE1 6TD. Telephone: Leicester (0533) 554100.

Manchester Museum, The University of Manchester, Oxford Road, Manchester M13 9PL. Telephone: 061-273 3333, extension 3101. Finds from the Peak District and Creswell Crags.

National Museum of Wales, Cathays Park, Cardiff CF1 3NP. Telephone: Cardiff (0222) 397951. Finds from Coygan Cave, Dyfed.

Passmore Edwards Museum, Romford Road, Stratford, London E15 4LZ. Telephone 01-519 4296 or 01-534 4545, extension 5670. Finds from Thames deposits at Ilford.

Sedgwick Museum of Geology, Department of Earth Sciences, Downing Street, Cambridge CB2 3EG. Telephone: Cambridge (0223) 333456. Much material, especially mounted skeleton of hippopotamus and remains of other mammals from the Ipswichian/Last Interglacial site of Barrington, Cambridgeshire, and mounted giant deer skeleton.

Somerset County Museum, Taunton Castle, Castle Green, Taunton, Somerset TA1 4AA. Telephone: Taunton (0823) 55504. Finds from Mendip Caves.

Stoke-on-Trent City Museum and Art Gallery, Broad Street, Hanley, Stoke-on-Trent, Staffordshire ST1 3DE. Telephone: Stoke-on-Trent (0782) 202173. Finds from Peak District Caves.

Swansea Museum, Victoria Road, Swansea, West Glamorgan SA1 1SN. Telephone: Swansea (0792) 53763. Finds from Gower Caves, South Wales.

Torquay Museum, 529 Babbacombe Road, Torquay, Devon TQ1 1HG. Telephone: Torquay (0803) 23975. Kent's Cavern material.

Wells Museum, 8 Cathedral Green, Wells, Somerset BA5 2UE. Telephone: Wells (0749) 73477. Material from Mendip Caves.

Woodspring Museum, Burlington Street, Weston-super-Mare, Avon BS23 1PR. Telephone: Weston-super-Mare (0934) 21028. Finds from Mendip caves.

Wookey Hole Caves Museum, Wookey Hole, Wells, Somerset BA5 1BB. Telephone: Wells (0749) 72243. Finds from caves; visit to Wookey Hole.

Yorkshire Museum, Museum Gardens, York YO1 2DR. Telephone: York (0904) 29745. Material of Ipswichian Interglacial age from Kirkdale Cave, North Yorkshire.

Index

Page numbers in italic refer to illustrations.